KEEPING BEES
AND MAKING HONEY

KEEPING BEES
AND MAKING HONEY

David and Charles

**ALISON BENJAMIN AND
BRIAN MCCALLUM**

Introduction

People are always surprised when we tell them we keep bees in our small city back yard. Is the reason for their incredulity the fact that beekeeping is a hobby associated with large country gardens and pastoral scenes? Our urban friends tend to think of a beekeeper as a middle-aged man sporting a beard and brown sandals. While the stereotype does exist, beekeeping is fast becoming a more popular pastime for younger people of both genders in towns and cities. It may be impractical for most of us to live the rural idyll, and indeed many of us would not want to, yet keeping bees can bring a little of the country-side into any garden.

At a time when a generation brought up on computers has less exposure to the outdoors – a condition dubbed 'nature-deficit disorder' – bees can also bring us closer to the world of flowers and trees whose nectar is the source of their honey.

Bees are fascinating, industrious creatures. They live together in a complex community and are incalculably important to the cycle of life, pollinating a large percentage of the crops we eat. When people ask – as they invariably do – what made us decide to keep bees, one of the main reasons is that we wanted to help the environment by providing a home for nature's master pollinator.

The growth of interest in beekeeping means there are many suppliers selling hives and equipment that make the hobby safe, easy and, with ready assembled components and a variety of labour-saving devices, far less time-consuming than it used to be. It is no longer the preserve of those with lots of time on their hands – even a high-flying city executive can keep bees nowadays. Like cats, bees do their own thing and require only occasional attention. The busiest time of the year is early summer, when you'll need to add extensions to the hive to give the queen bee enough room to lay her eggs and the workers sufficient space to store their honey. But even then, they will only take up a few hours of your week. In the winter, you need do nothing except occasionally check that the hive hasn't fallen over and the bees' honey stores haven't become seriously depleted – and preparing for the year ahead is also a good idea.

Most beginners find that it is worth joining a local beekeeping association, where they can pick up tips from experienced beekeepers keen to share their knowledge. But a word of warning: there is a saying that if you get three beekeepers together you will get four opinions on how best to manage your colony, and this is not far from the truth. The art of beekeeping is constantly evolving and beekeepers continue to invent their own ways of managing their colonies. Therefore there is no definitive guide on how to keep bees.

The goal of many beekeepers is to get as much honey as possible from their bees and they measure their success based on the quantity harvested at the end of the season. This, however, is not our goal, nor is it the purpose of this book. What we hope to do is introduce you to the basic principles of modern beekeeping and enjoying the fruits of the bees' labour. As well as honey, bees produce wax, which for centuries has been highly prized in candles, polish and cosmetics. With a little practice you will be able to make your own candles, furniture polish and even hand and body lotions.

Beekeeping is endlessly fascinating, from the variety of honey the bees produce – no two harvests ever seem to taste the same, even from the same hive – to the different behaviour displayed by different colonies. You may find that before long what started as a passing interest becomes something of an obsession. And, as with anything else, the more deeply you delve into it the more complicated it becomes.

However, another saying among the beekeeping fraternity is that bees don't read books and there are many instances of bees behaving in ways that surprise even the most experienced apiarists. So don't be concerned if your bees do something unexpected. We can only assume that they know best – after all, they have been looking after themselves for millennia without any help from us.

Above all, this book aims to illustrate the huge enjoyment you will derive from having one of nature's most wondrous creatures residing in your garden. We really hope it will whet your appetite for beekeeping and encourage and inspire you to bring bees into your life, wherever you live.

Alison Benjamin and Brian McCallum

Understanding bees

The bee is one of nature's wonders. As well as being its most effective pollinator, honey bees produce delicious and healthy food that humans have been eating for centuries. While they can't be tamed, they can be given shelter and protection from the rain and cold. In return for providing them with a home, we can find out about the intricate workings of a colony, discover the joyful art of beekeeping and enjoy some of the bee's sweet-tasting food stores. In addition, a hive is an attractive feature in any garden or outside space.

History

Honey bees and humans

Humans have revered bees for centuries. Not only did they supply us with one of the first natural sweeteners, versatile medicines and precious commodities, but the workings of the hive and its residents' industrious behaviour have provided endless fascination as well as countless lessons. Political theorists have based numerous models of society on bee colonies, business-men have drawn inspiration from their division of labour to develop management practices and the construction of their hexagonal honeycomb home has taught architects about the principles of design.

We know that beekeeping had begun by 30BC, because the Roman poet Virgil included a practical treatise on the art in his *Georgics*, in which he vividly describes the workings of a hive:

Aware that winter is coming, they use the summer days
For work, and put their winnings into a common pool.
Some are employed in getting food, and by fixed agreement
Work on the fields; some stay within their fenced abode,
With tear of daffodil and gummy resin of tree-bark
Laying their first foundation of the honeycomb, then hanging
The stickfast wax: others bring up the young bees, the hope
Of their people: others press
The pure honey and cram the cells with that crystal nectar.
Some allotted the duty of sentry-go at the gates….
Relieve incoming bees of their burden, or closing ranks
Shoo the drones – that work-shy gang – away from the bee-folds.
The work goes on like wildfire, the honey smells of thyme.

No wonder Virgil describes the hive as a 'tiny republic'. No wonder, also, that throughout history, from Aristotle and Plato to Erasmus and Marx, the bee colony has been held up in support of wildly different forms of society, from absolute monarchy to republicanism and socialist utopia.

The bees' relentless toil has been proffered as an example for humans to emulate throughout the ages. The expressions 'hive of activity' and 'busy bee' have entered the English language, while the name of the non-working male, referred to as the 'lazy yawning drone' by Shakespeare in *Henry V*, is now used to describe a slothful individual. It is no coincidence that in the books of PG Wode-house the Mayfair establishment where fashionable young men waste their days is called the Drones Club.

Right *A 15th-century portrait of Virgil (70–19BC) writing the 'Life of Bees' from* **The Georgics.**

The worker bee's industriousness was also embraced by the early colonizers of North America, who brought the honey bee with them from Europe. While peace has its olive branch, industry has its beehive, and US companies were quick to choose the bee and its home for their logos. The midwest state of Utah, whose motto is industry, is known as the Beehive State.

Despite their heavy workload and their chastity – save for the queen and a few drones – bees have also been closely associated with love and sex. Golden, dripping honey has symbolized the deliciousness sweetness of both love and lust since biblical times: the Old Testament *Song of Songs* contains the lines

Thy lips, O my bride as the honeycomb
Honey and milk are under thy tongue.

And the expression 'the birds and the bees' is still a euphemism for the mechanics of sexual reproduction.

For centuries, however, we got the gender of the big bee in the hive wrong. Just as it was assumed that human rulers had to be male, so the same prejudice was applied to the beehive: 'They have a king,' wrote Shakespeare in *Henry V*. It wasn't until English bee expert Charles Butler published his famous thesis, *The Feminine Monarchy,* in 1623, that the hive was conceived of as a matriarchal kingdom. Butler's ideas were confirmed a century later by Dutch scientist Jan Swammerdam in his *Book of Nature.*

For as long as humans have been studying bees we have also been utilizing the products of their labour. As much as 3,500 years ago the ancient Egyptians were making beeswax candles to provide artificial light. In the 12th century candlemaking became a profession practised by wax chandlers and there is still a

Worshipful Company of Wax Chandlers in the City of London. Their main customer was the church; wherever Christianity spread in Europe, so did beekeeping and candlemaking. It is said that in Britain beekeeping never recovered after the Reformation which destroyed the monasteries, as monks and nuns were among the most enthusiastic apiarists. The bee then became associated with the Catholic church and from the 17th century was an official symbol of papacy. Not until 1900 were Catholic churches allowed to use non-beeswax candles.

Honey, beeswax and mead also had close ties with monarchs and landowners, who would accept them as payment of taxes and tithes. The practice stretches all the way from the ancient Aztec rulers of Mexico to feudal Welsh kings. The Greeks minted coins with bees on them and taxes were later demanded on bee products. In 16th-century France,

beekeepers destroyed their hives rather than pay higher taxes.

In Greece and Egypt early hives were made of pottery and shaped like giant thimbles. The Romans used hives made from all manner of things – logs, bricks and even dung. In Europe, from the Middle Ages, wicker skeps were the predominant type. As they would rot if exposed to bad weather, special holes were made in walls in Britain and Ireland to shelter them. These 'bee boles' can still be seen in grand 17th- and 18th-century country houses, such as Packwood House in Warwickshire. It was not until the invention of movable frame hives around the mid-1800s that modern beekeeping began.

Below *A 19th-century engraving entitled a 'study of different bees'.*

HONEY BEE. 1.Worker. 2.Male. 3.Queen. 4.5. COMMON HUMBLE BEE. LAPIDARY BEE. 6.Male. 7.Female. 8.MOSS or CARDER BEE. 9.DONOVAN'S HUMBLE BEE. 10.HARRIS' HUMBLE BEE. FALSE HUMBLE BEES. 11.Apathus Vestalis. 12.Apathus Rupestris.

Bee species

Entomologists, with their increasingly sophisticated technology, have identified more than 20,000 species of bee worldwide. The bees we are most likely to see in our fields and gardens represent just a handful of these.

Solitary bees

The vast majority of bees are not into communal living. They have flying lives of no more than six to eight weeks and spend most the year as pupae in a cocoon. All the females of these species are fertile, once mated, a female lays around eight female and one male egg, each in its own cell, in a small burrow or tunnel. The burrows can commonly be found in the ground, soft mortar, wood or the hollows of reeds and canes. The larvae spend most of the year developing in their individual cells; they emerge only as their food source is about to blossom. The first bee to emerge is the male, which quickly mates with any females he may find. He then dies, leaving the female to look for a new nesting site to continue the life cycle.

Unlike the bumblebee and the honey bee, solitary bees usually feed on one plant, so when the plant's season ends the bee's food supply dries up and it dies. As a result, different species are seen at different times of the year.

You can encourage solitary bees to nest in your garden by installing a nesting site comprising a bundle of thin tubes. These can be made from 15cm (6in) lengths of 6mm (¼in) bore bamboo canes contained in a length of guttering. Attach this to a fence or wall in a sunny position around head height and with luck the bees will nest there. Some garden centres now sell ready-made bee boxes (below left) alongside their bird box selection.

Bumblebees

The bumblebee (*Bombus*), of which there are around 250 species worldwide, is usually larger and rounder than other bees. It has a thick furry coat, which allows it to fly in colder weather, so in temperate climates it is generally the first bee to be seen in spring. Bumblebees live in small colonies, with populations numbering from 50 to 400, and a queen who produces female worker bees.

Only the queen bumblebee survives the winter by hibernating in the undergrowth. In the spring she emerges and searches for a suitable nesting site in which to lay her eggs and build her colony. This nesting site will generally be below ground level in old mouse holes, burrows in the undergrowth or even cracks of concrete in the street. The queen quickly builds the wax cells to accommodate her sterile, female offspring and as the population grows these worker bees gather the pollen and nectar for the colony.

Near the end of the summer, the colony raises virgin queens and males. They both leave the nest to find a mate and while the males die after mating, the successful queens will hibernate soon afterwards. The vacated colony will die as winter approaches, then spring will see the cycle repeat itself with the next generation.

The buzz of bumblebees gathering nectar is a familiar summertime sound. But if the alarming decline in numbers is not halted it could become a thing of the past. In Britain alone, three species are now extinct and nine of the remaining 22 are endangered. Intensive farming methods have destroyed hedgerows and meadows (the bees' favoured nesting grounds) and increased use of insecticides has poisoned them.

You can make a patch of your garden more bumblebee-friendly by growing flowers that are rich in pollen and nectar and easily accessible to passing bees, and by having some plants that flower throughout the year. Providing a haven for the queen in winter by leaving areas of your garden untouched, especially with soft earth and coverings of leaves, can also help bumblebees to thrive. In spring, to encourage them to nest, put some dried moss or hamster bedding in an earthenware pot, then place the pot upside down in a hole in the ground. Cover the hole with a piece of slate or tile to stop the rain getting in but leave enough room for the queen bee to get under the tile. Hopefully she will find the pot and make it her home.

Honey bees

There are four main species of honey bee (*Apis*), which vary in size and characteristics, but all live in large colonies and store honey for winter supplies.

The species whose honey humans have harvested for centuries is the western or European honey bee, *Apis mellifera*, and it is this species that we will be concentrating on in the following chapters. So productive is this bee that it has been introduced all over the world to make honey and pollinate crops.

Honey bees are very well adapted to their role, with their long tongues (proboscis)

designed to suck up nectar and their furry bodies to trap pollen. They are typically 15mm (just over ½in) long and live in colonies with populations measured in tens of thousands. They use wax, secreted from wax glands on the underside of their abdomen, to build the complex structures of connecting hexagonal cylinders we call honeycomb, in their nests. These nests are usually the size of a football but can be larger if space permits.

All bees are important to the planet because they pollinate much of the flora. Whether you keep honey bees in a hive or encourage their cousins to nest in your garden you will be helping to bring back an important insect that has been in decline.

Wasps

Many people group bees and wasps together in their minds because they look similar, buzz and can sting. But though distant cousins they are very different creatures. Wasps are omnivores and will kill for a meal, whereas bees eat only nectar and pollen and are quite passive. It is the wasp that hangs around your plate when you are having a barbecue, searching for any scraps of meat or fish, or diving into the jam pot while you're enjoying scones and cream in the sunshine. Wasps are persistent and a nuisance. Bees, on the other hand, are not interested in your food.

Wasps don't have wax glands, so make their nest from a mulch of wood fibres akin to papier mâché. They can sting more than once, whereas honey bees die after one sting, making them less likely to attack.

Though wasps get a bad press, and people often ask 'What are they for?', they do have a useful role to play in early summer, eating aphids, small flies and any decaying matter that is lying around in the garden.

PROPOLIS

Propolis is a resinous substance that bees collect from the barks of many trees and plants. It is sticky at 35°C (95°F) – hive temperature. It is usually brown or yellowish brown in colour. Poplar, birch, beech, horse chestnut and sunflowers are common sources.

The resin is used by the bees:
- as a glue to bind the nest together and seal up holes and gaps
- as an antiseptic when lining the cells to protect from moulds and infections
- to mummify dead intruders such as mice which are too big to be removed from the hive.

It is used by humans:
- to varnish the woodwork of hives
- as a health product with antiseptic properties, often sprinkled over breakfast cereal.

The great violin maker Stradivarius also favoured it to varnish his instruments.

The caste system

Within a colony of honey bees there are three castes: queen, worker and drone. Each has a distinct function to perform within the colony.

The worker bee

During the summer, when there is ample food around, a healthy colony of bees will typically be made up of one queen bee, a hundred or so male drone bees and 50,000 female worker bees. It is the workers that we see dancing around the flora of our gardens.

All matters relating to the wellbeing of the colony, except for egg-laying, are handled by the worker bees. Their mission in life is the survival of the colony. In pursuit of this goal, they work together through a complex communication system. Their duties around the hive change with age (see page 17) and they have wax glands to build the honeycomb.

Workers communicate to each other by touch, smell and vibrations. The Nasonov gland at the end of the abdomen emits a pheromone (scent) that binds the colony together. A worker bee will often be seen standing still at the hive entrance, with her bottom in the air, beating her wings rapidly, wafting the scent into the air to guide her fellow workers back to the hive.

The drone bee

The drones are the males of the colony. They are physically larger than the workers, squarer in shape, with big eyes. Their sting has been replaced with the male sexual organ, and they don't have wax glands. Their main role is to search out, find and mate with a queen.

The colony will produce a few hundred drones in the spring in anticipation of the needs of a virgin queen. Interestingly, the drones are hatched from unfertilized eggs, which means they don't have a father. Since they are larger than the workers, their cells are larger and can be seen in the hive quite easily. When they are mature they will take to the sky to find a receptive queen with whom to mate. The mating process actually kills the drone, but by that time he has ensured that his genes are passed down to future generations.

Above *A drone bee takes to the sky to find and mate with a queen.*

The queen bee

We give the queen bee a royal title but in fact she has no real powers over the colony. She doesn't rule or command them, but without her the colony would die. Her role is to lay eggs and she does this continuously in the dark for the duration of her life, reaching 2,000 a day at her peak. She even has a group of attentive maids who feed, clean and clear up after her so that she doesn't need to disrupt her egg-laying with such trivial matters.

The queen emits a variety of pheromones which bind the colony together. The exact purpose of all these scents isn't clear, but we know that one of them reassures the colony.

Right *The queen bee surrounded in the hive by an attentive team of workers who feed, clean and clear up after her.*

The anatomy of a honey bee

The Nasonov gland emits another pheromone that says, 'Here we are' and is used when the bees are swarming or to guide a forager home.

The wax squeezes out as a liquid from the wax glands on the underside of the abdomen, where it solidifies into tiny scales of wax.

The hypopharyngeal and mandibular glands, inside the head, are used to make royal jelly (see page 16).

The sting and the associated poison are used in defence. The sting is barbed, so that it can attach itself to the fibres of the recipient. When the bee pulls away, her insides stay attached to the lodged sting, subsequently tearing her in two. When a bee stings she emits a pheromone which alerts her sisters to the danger and directs them to sting in the same location.

A honey bee has five eyes: three simple eyes at the top of the head, which she uses for orientation, and two complex eyes, which are for seeing. A bee sees colour and shapes: it doesn't see slow-moving objects well. Fast, jerky movements will attract much more attention – something to keep in mind when working around bees.

Abdomen

Thorax

Head

The antennae are used for smelling and are very sensitive to variations in odour.

The two hind legs hold the collected pollen on the corbicula (pollen basket).

The mouth has two mandibles which act like teeth and work like pinchers. The bee uses them to manipulate wax, to collect propolis and to bite.

The four wings and six legs are attached to the thorax – the powerhouse of the bee. The smaller hind wings are hinged to the more powerful front wings, allowing the hind wings to fold on to the body. This gives the bee more room to move around the crowded colony and get into small flowers for nectar. The wings can also be rubbed against each other to make a sound that is used in communication.

The honey stomach holds the nectar after the bees have collected it from the flowers. Enzymes act on the nectar, which starts the process of honey making.

The two front legs have antenna cleaners.

The two middle legs clean the wings, brush the pollen on to the corbicula and take the wax from the wax pockets on the underside of the abdomen.

The proboscis is the tongue, which acts like a straw to suck the nectar into the honey sac. It is up to 7mm (1/4in) long, good for getting deep into the flowers.

The birth cycle

The queen bee, the mother of all the other bees in the hive, continually searches for clean cells in which to lay her eggs. She backs her way into one of the cells and attaches an egg to the back wall with a type of glue. She then moves to the next available cell to lay her next egg. She may do this 2,000 times a day.

Each egg, reminiscent of a small grain of rice, hangs horizontally from the cell wall. It takes three days for a larva to hatch from this egg and when it does, like all newborns, it is hungry and needs food in order to grow. Worker bees are on hand to feed the larva with the nutrient-rich substance we call royal jelly. This jelly – secreted by the workers as part of their saliva – is fed to all larvae for the first three days, after which those that have not been chosen to become queens are switched to a mixture of pollen and honey.

The larva now goes through a series of instars (moulting of the outer skin). With each stage it grows bigger and has to curl up into a 'C' shape to fit into its cell. After six days the pearly white larva straightens itself out and stops eating. The attending bees notice this change of behaviour and build a cover over the cell, using a mixture of wax and propolis to enclose the larva for the next stage.

During the next 12 days the covered larva pupates, spinning a cocoon and changing from a grub into an insect. When this metamorphosis is complete it eats its way through the cell cap, emerging as a young honey bee.

For a worker this process takes 21 days, for a drone 24. The queen, in her specially enlarged cell, develops more quickly and emerges from her cocoon after only 16 days.

Beekeepers soon learn to differentiate between each of the brood stages. Easiest to see is the capped pupa cell, but with experience you will learn to recognize the fat white larvae and the tiny eggs. This skill will become important when you need to check the health of the colony.

The area within the hive that contains the eggs, larvae and pupae is called the brood nest and is oval in shape. This area increases as the queen lays more eggs in the warmer months but the overall shape stays the same.

ROYAL JELLY

Royal jelly is a white, jelly-like, protein-rich substance made by the bees from the hypopharyngeal glands located in their heads. In addition to using it to nourish the young larvae, attendant worker bees also feed royal jelly to an adult laying queen.

Royal jelly is thought to have rejuvenating powers for humans and is a much sought-after ingredient in cosmetics. But realistically, as an amateur beekeeper with a few hives, you are unlikely to be able to harvest it.

Left *The birth cycle of the honey bee.*

The life cycle

Workers

The duties that a worker bee carries out throughout her life change with age and physical maturity. A newly emerged bee is not fully developed – her sting and wax glands, for example, have yet to mature.

1. She begins her working life as **a cleaner** straight after birth. She will inspect, clean and polish cells for the queen to lay in.

2. A few days later her task is to **feed the older larvae** with pollen and nectar. When her own hypopharyngeal and mandibular glands have developed, she can start to feed the younger larvae with royal jelly.

3. The wax glands develop next, so she can begin to **cap the larvae** and honey cells, and use the wax to repair and build new comb.

4. Two weeks old, and she can **collect nectar and pollen from returning foraging bees** and place this food in the store cells.

5. 18 days old – three days before she begins to fly outside – she will start to **guard the entrance to the hive**. Her sting has fully developed by now and her mandibles are strong. She spends time at the entrance smelling the returning bees to make sure they are from her colony. If they are not residents, like a bouncer at a nightclub she bars their entry. If the chancer won't take no for an answer, she will resort to using her sting to kill the intruder, but will die in the process.

6. During these first three weeks in the hive, the worker's normal tasks may be interrupted by more urgent ones, such as **ventilation**: flapping the wings rapidly circulates the air and helps to regulate the temperature of the hive. A bee also fans nectar to evaporate the water content, turning the nectar into honey.

7. Three weeks old – she is ready for **foraging**. She will take a few orientating flights to get her bearings and then spend the rest of her life – a further three weeks – collecting nectar, pollen, water and propolis.

This is draining work and is fraught with dangers. She may fall prey to spiders and birds, her wings get battered from hitting the flora and she will travel miles carrying a load of pollen and nectar weighing as much as she does. In theory a bee can live for several years, but in practice, six weeks after being born, most workers have died of exhaustion.

However, a worker who is lucky enough to be born in late summer could be one of the 10,000 bees who live through the winter, huddled together eating the stored honey. Her luck will come to an end when the foraging season returns in the spring and she is among the first out to search for new food.

Drones

The role of the drones, the male of the species, is to mate with a queen. When they are not out flying looking for a queen on her mating flight, they spend their time lazing around, eating honey and making a mess of the hive. They rely on the colony for their food and shelter, and though they may help with ventilation they do little else around the home. They can live like this for most of the summer and, unlike their sisters, they can migrate between colonies while they look for their prize. They mate in flight, but this is the last action of their lives. As the drone dismounts from the queen his sexual organs are ripped from his body and he falls, dying, to the earth. Any drone who fails to mate and is still in the hive towards the end of the summer has outlived his welcome. The colony can ill afford to waste its stores of honey on non-productive members, so these surviving drones will be evicted by their sisters and, once outside the hive, will soon starve to death.

The queen

Unlike other bees, a queen can live for years. She spends practically all of that time in the darkness of the hive, venturing out on only two occasions: once at the beginning of her life to mate and later to swarm.

Any fertilized egg can become a queen if the colony decides it needs to raise a new one. The chosen egg's cell is enlarged and built to hang vertically (worker cells lie horizontally). The workers feed the growing larva intensively on royal jelly (every five minutes for 5 1/2 days), which stimulates it to develop into a queen. The metamorphosis takes 16 days and, after a few days of acclimatizing, the new virgin queen will leave the hive to mate.

On this nuptial flight the queen's pheromones attract the drone bees and she mates with up to 12 of them until her spermatheca (the sac that holds the sperm) is full. On returning to the hive she begins her life as an egg-layer. When she eventually starts to lose her egg-laying abilities due to old age or illness her pheromones change and this signals the workers to raise a new queen. Sixteen days later the new queen will kill her mother and take over the hive, a process called supersedure.

What to Consider

Before deciding to keep bees you need first to be clear about your aims. Ask yourself whether you are interested in profit or whether you want to keep them purely for the pleasure of doing so. You should also weigh up how much time and space you have at your disposal. Little of either need not preclude you from keeping bees, but it will mean that owning more than one or two hives is out of the question and it is unlikely you will produce enough honey to make your hobby pay for itself.

Space

Keeping bees is traditionally thought of as a rural hobby, the preserve of people with large country gardens or acres of land. But it is becoming more popular among urban residents and suburb dwellers. One reason for this is that a beehive doesn't actually take up much room – it can fit on to a standard paving stone – so you don't need a large garden to accommodate it. Hives can be kept in the smallest of back yards, on roof gardens and even on balconies (right). More important than a large space in which to put the hive is adequate storage for all the paraphernalia that goes with beekeeping (see page 42). This might be a small outside shed, a large cupboard, or space in the basement.

The fact is that urban bees do very well, possibly because there is more biodiversity in towns than in the countryside these days. It is said that, because of our increasingly industrialized agricultural practices, 0.2 hectares (½ acre) of overgrown churchyard is home to a greater variety of plants and wildlife than 16 hectares (40 acres) of monoculture arable land. In addition, city gardens and their flora are as diverse as their owners and produce an abundant variety of year-round forage. Add to this mixture the parks, tree-lined roads and undisturbed railway sidings and you will find that bees are more than happy with an urban environment.

Locating the hive

The location of your hive is the first thing to think about. In this respect, smaller gardens present some issues that larger properties generally don't, such as the hive's proximity to your and your neighbour's backdoor, but the principle for finding the best place to put your hive is the same for everyone. You want comfort for yourselves and the bees, so choose a spot that is dry, sheltered, undisturbed and far enough away from people that neither you nor the bees are interrupted in your daily business.

When they leave the hive to forage, bees naturally fly upwards in a series of increasingly large spirals. This orientates them and imprints the mental aerial map they need for their return journey. When they return they are heavy with their booty and will fly directly

back to the hive. If you are putting a hive close to where people walk you can encourage the bees to fly above head height and down to the hive by placing some 1.8m (6ft) hedging or fencing around the hive. The bees will soon learn that it is easier to fly directly down to the hive rather than approaching at a low level and then having to climb to negotiate a high barrier with a heavy load.

Bees like a bit of morning sunshine, so if possible site the hive in an area that catches the early rays. This warms the hive and lets the bees know that the day has begun: getting them out of the hive early will increase their foraging time, which should result in more honey at the end of the season.

Like most creatures, bees like to be sheltered from the elements. The hive itself will protect them from most of what nature throws at them, but don't put it in a windy spot and don't have the entrance facing the

> **Bees like a bit of morning sunshine, so if possible site the hive in an area that catches the early rays.**

prevailing wind. Bees are strong and can fly in windy conditions, but no one likes a draught.

Cold

Although bees regulate the temperature inside the hive, try to keep it somewhere that gets neither too hot nor too cold. If your chosen site is in the sun all day, give the hive some shade either by planting some foliage or by putting up a fence. Avoid sunken areas, such as a hollow or dip, where cold air can get trapped – this will mean the colony has to work harder at keeping the brood at the required 35°C (95°F).

A hedge or some form of wind break is desirable if the hive is to be in an open spot. If the area is prone to damp, place the hive on a high stand (see page 37). Bees don't mind the cold, but they dislike the damp. Although shading from trees is helpful during the summer to prevent the hive overheating, the branches mustn't droop so close to the hive that they will bang against it when the wind blows. This will disturb the bees, and in a storm branches could even fall on the hive and destroy it.

Water

If the area you live in doesn't have a natural source, you will need to provide some fresh water. This can be from a water feature in your garden or a pot filled with sodden peat. Once the bees have found the supply they will be frequent visitors, but as they can't swim you need to make sure they have something to stand on while they drink – otherwise they are likely to drown. Floating sticks will suffice. Keep the supply of water a few metres (yards) from the hive so the bees can defecate without contaminating the water supply. They are very clean creatures and only relieve themselves outside the hive, even in winter when they have to wait for a warm day before they can venture out.

A bee shed

You can even keep bees in a garden shed. Bee houses, as they are known, are especially common in northern Europe. They are practical, as the equipment, clothing and sundries can all be kept inside the shed too. Just make a slit in the side wall of the shed and place the hive on a shelf with the entrance facing the slit. On the outside, attach a piece of wood – 30cm x 9cm x 2.5cm (12in x 3in x 1in) is fine – as a landing board to aid the bees' return. Depending on the size of the shed, a number of hives can be kept this way side by side – if you have more than one, paint

Below Left If the hive is in a windy location, you may need to put a weight on the roof to keep it in place.

Below A bee shed is a good way to keep a number of hives. If you paint the landing boards different colours it helps the bees to find their way home.

each of the landing boards a different colour to help the bees return to the correct hive. The downside of this approach is that you forgo the aesthetic pleasure of having a hive to look at in your garden.

Two hives

It is quite common for the novice beekeeper to become so enthralled with the hobby that they soon think about increasing their apiary,

so it is worth considering a site that may have the space to accommodate more than one hive. All the same rules apply, but you should angle the second hive's entrance to face in a slightly different direction, so that the bees know which hive to return to, and place it at least 1m (3ft) away from the first.

Out apiary

If you don't have anywhere suitable for

keeping bees, consider asking a local land-owner if you can keep a hive on their land in return for some honey. Arable or animal farms, part of a cemetery that is not open to the public, the grounds of a retirement home, public allotments, secluded areas within local parks, or brown field sites to which the public does not have access, such as overgrown land around disused warehouses or factories could all make suitable locations for a hive or two.

In addition to supplying jars from the honey harvest, you could also offer to give a talk about beekeeping to residents of the retirement home, the church congregation or users of the park.

If you plan to keep a hive in a location far from your house, there are additional factors to consider. You want it to be easily accessible, as there will be times when you have to take extra equipment to it and other times when the honey is ready for collecting that you will have to carry a fairly heavy load back. So don't put the hive anywhere that forces you to walk a long way, or where there are narrow gates and openings to negotiate when your hands are full.

Keep the hive away from public footpaths and thoroughfares. As well as the risk of people being stung, hives can sometimes be the target of vandalism, abuse and theft.

tip
If the area you live in doesn't have a natural source of water, you will need to provide fresh water.

Children

There seems to be something captivating about a hive. The foraging workers returning and telling the gang their news, or the queen steadily walking around the frames looking for a clean chamber in which to lay her eggs, can keep most of us intrigued, whatever our age. In addition, many children have been brought up on stories about a certain bear and his love of honey.

Bees are not, of course, pets in the traditional sense. You can't take them for walks, stroke them or cuddle them, but then again you don't have the chore of cleaning out their home as you do with a rabbit or guinea pig, and you do get that sweet honey as a treat.

Bees can become part of the family and in today's society, they are a perfect antidote to 'nature-deficit disorder'. What better way for children and their parents to learn about the workings of nature together?

The worry of being stung is always present with bees, especially when there are children around. But some simple precautions will keep everyone safe and happy. Many bee suits come in children's sizes, so they can easily join in with the hive inspections. Children are often eager to ask questions and learn what's going on, and there is a good chance they will spot the queen on a frame before you do.

> The worry of being stung is always present, especially when there are children around. But some simple precautions will keep everyone safe and happy.

Honey-harvesting time is particularly fun for children: sticky, sweet and messy, just the sort of thing they love. They'll especially enjoy spinning the extractor to get the honey off the frames (see page 84). Many schools have harvest festivals around this time, so children can also have the thrill of contributing a jar of home-grown honey.

And it's not just about honey. Candlemaking is a craft that you and your children could learn and enjoy together (see page 108).

Pets

Many people cite their pets as a reason why they can't keep bees. You may be worried about cats or dogs getting stung, but cats at least are more sensible than you might give them credit for. Although the roof of an empty hive – before the bees arrive – may be a favourite sun trap, they will soon learn to steer well clear once the bees move in. Dogs can be a bit more inquisitive, so be sure to shut them in the house when the hive is open for inspection.

Horses can kick hives over, accidentally or on purpose, so try not to have bees in the same paddock. If this is unavoidable, protect the hive with some sort of barrier.

As you will find out, bees are generally passive creatures who are content getting on with their own lives and happy to coexist with you and your pets.

Below *Children are fascinated by bees, and will enjoy taking part in hive inspections and maintenance. Bee suits come in children's sizes, so beekeeping can be a safe, family experience.*

Routine

Beekeeping isn't a labour-intensive activity. For the most part bees look after themselves and our role is to make sure the colony is kept healthy and has the space it requires. After all, bees have been getting by without us for millions of years so you will find that, with little interference, they will happily produce all they need for their own purposes – and a bit of extra honey for you.

The work involved varies with the seasons. In winter there is little need to check on the hive. The occasional visit just to make sure that it has not been knocked over or a tree branch hasn't fallen on it is about all that is required. There will be some work to be done away from the hive such as making the frames and supers (see page 38), but this doesn't take long and gets quicker with practice.

During the warmer months the workload increases and at certain times weekly visits are called for, but even so one colony shouldn't take up much more than a couple of hours per week in the busiest period and an hour per month during the winter.

It is useful to keep a notebook and enter everything you have done or seen at the hive. Especially if you have more than one hive, it is surprisingly easy to forget what you did when and at each one. A notebook also helps you plan ahead and schedule the next visit.

Going away

One advantage of bees' requiring only the occasional visit is that you don't have to rely on your neighbours' goodwill while you are on holiday. Even if you have to go away at the busiest period, in early summer, don't worry. With some planning and preparation, such as

adding a super or two in case the bees run out of space (see page 62), you can safely leave them to get on with things.

Cost

Small-scale beekeeping doesn't need to be expensive. The initial nucleus of 10,000 bees, a hive and all its components, bee suits, smoker

Above *It is useful to have a basket or container for all of your essential beekeeping tools, including a smoker, fuel and gloves.*

and other equipment are the biggest outlay. Thereafter, the running costs are fairly low: a few pounds of sugar, pots, jars and labels are the only likely extras. There are many suppliers selling a whole range of gadgets and mechanized apparatus who will gladly take your money, but it is not necessary to purchase absolutely everything. More pricey items such as honey extractors can be shared or hired from local bee associations.

Shop around for your first hive and equipment. Prices vary enormously, but as with most things quality merchandise lasts longer, so buy the best you can afford. Specifically, go for a cedarwood hive if you can (see page 33). There is a second-hand market, but make sure you are buying from a responsible outlet, as disease can be transmitted through used equipment. Make sure everything has been cleaned, scorched and sterilized. Your local bee association will recommend responsible resellers in your area.

As you will see on page 32, hives are really just modified four-sided boxes. Hives and frames can be bought either made up or as a flat pack. Flat pack is cheaper and is delivered more quickly (which can prove crucial) if you're buying mail order, but it takes time to assemble and if you do not know one end of a hammer from the other it can be daunting at first. If you like carpentry, you can construct a hive from scratch using drawings and plans readily available on the internet.

Neighbours

Not everyone will be as happy as you are with the idea of a beehive in their vicinity. It is only courteous to alert your neighbours to the new arrivals who will shortly be moving in. Although there may be hostility initially, it should take only a few weeks for everyone to realize that bees keep themselves to themselves and are not interested in humans. Try to allay your neighbours' fears and reservations by highlighting the benefits of keeping bees. Inform them that: bees are the best pollinators around, so trees, flowers and fruit in the neighbourhood will flourish; they give us gorgeous honey – and you will gladly give the neighbours a jar or two; and they are a wonderful way to see nature at work. If anyone in the area suffers from hay fever, impress on them the medicinal benefits of eating locally produced honey.

The major fear will be about being stung. Reassure them that bees are not aggressive, and because they eat only nectar and pollen, they are not interested in our food – unlike their carnivorous cousins the wasps. Bees are also quiet, don't smell and, despite some popular misconceptions don't chase after people in large, killer swarms.

BEE STINGS

Bees are not angry, nasty or hostile creatures and they are not sting-happy, because the action of stinging kills them. But they are defensive, recognizing when they are under threat and responding accordingly. As we are sometimes intruding into their home it is inevitable that they will try to sting. Most people react to a sting with a localized swelling followed by a couple of days of itchiness. The swelling can be large and red but this is completely normal. Some people, though, have a serious allergic reaction to bee venom (in the same way that some people are allergic to nuts) which can result in a life-threatening anaphylactic shock. If you are at all concerned about the reaction that you or your neighbours may have, seek professional medical advice.

Local bylaws

Check with your local council in case there are any ancient laws that forbid beekeeping in your area. This is extremely rare but not unknown.

Above It is a good idea to tell your neighbours that you will be keeping bees once you have decided where you are going to locate your hive.

tip
Try to allay your neighbours' fears and reservations by highlighting the benefits of keeping bees.

Housing your bees

How your bees are housed depends on considerations such as whether you want an aesthetically pleasing hive or one that is easy to use and how much money you want to spend. The initial cost of a good hive and its components will be your major outlay but they will last a long time if they are looked after well. There is a wide range of hives on the market but most follow the single-walled box model. During the summer when the flowers are in bloom you will increase the height of the hive to give the bees additional space to store their honey. It is not uncommon for a hive to stand 1.5m (5ft) or more tall by the end of the season.

How a hive works

As we have seen, bees are social insects living together in a colony with a queen, thousands of workers and a few drones. The colony itself can be thought of as a living organism and each individual bee as a component part whose job is to keep the organism alive.

In the wild, bees live happily in dark, dry cavities with small defendable entrances. Inside this cavity they build their nest from self-generated wax. Workers shape the wax into connecting hexagonal cells that protrude from either side of a vertical sheet hanging from the roof. As a colony grows, the workers hang more vertical sheets – a large colony may have eight sheets, with the central ones being the longest and widest. The cells on these sheets are used for birth chambers, honey and pollen stores and resting rooms.

The distance between each sheet is constant and uniform and is about 9mm (⅓in) for the western honey bee. This distance is just enough for the bees to crawl around in and pass each other without getting crushed, but not so far apart that they lose the important bodily contact. This distance is important for beekeepers and is known as 'bee space'. More on this later.

Prior to the invention of the modern-day hive, when colonies were kept in upturned baskets called skeps, a colony was destroyed when the honey was collected. This wasn't good for the bees or the beekeeper.

Thanks to the work of some eminent entomologists in the mid 19th century, a better system was developed. This exploited the principle of the bees' natural behaviour of building sheets of comb. A number of frames were housed in a box and the bees used these frames to build their vertical sheets of wax comb. When a frame was filled they moved on to the next one. More frames could be added when needed. The advantage of this system was that the frame could be lifted out of the box without the colony being destroyed. This was called the 'movable frame hive' and although it was a combination of many people's ideas it has been attributed to the Reverend LL Langstroth, who patented it. His eponymous hive is widely used today, most notably in the USA.

Reputedly, the first box that was used to hold these removable frames was an empty champagne crate, and today's hives haven't changed that much. What you get when you buy a hive is a number of roofless and floorless boxes that sit on top of each other, each capable of holding ten or so frames. The boxes come in two depths, deep and shallow, usually known as brood boxes and supers. Brood (deep) boxes are so called because this is generally where the brood (developing larvae) is housed while the supers – Latin for on top – are above the brood nest and are where the bees store their honey. The reason they are more shallow is so that they are not too heavy for us to lift when they are full of honey. The bees don't mind the size of the boxes or the frames.

A hive also has a roof, a floor and a stand. The tower they make has a footprint of around 60cm (2ft) square and a height ranging from 60cm–1.8m (2–6ft) depending on the season – you need to add extra floorless super boxes to save the incoming nectar as the summer progresses.

The art of beekeeping lies in knowing how many frames and boxes to give the hive, when to add and remove them, and recognizing what is on the frames. We will be explaining this in the following chapters.

Right *An illustration showing the different components of the hive, and how it works.*

Below *Moveable frame hives allow you to remove the frames of honey from the hive without damaging the colony.*

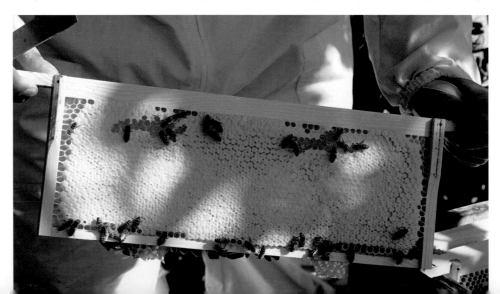

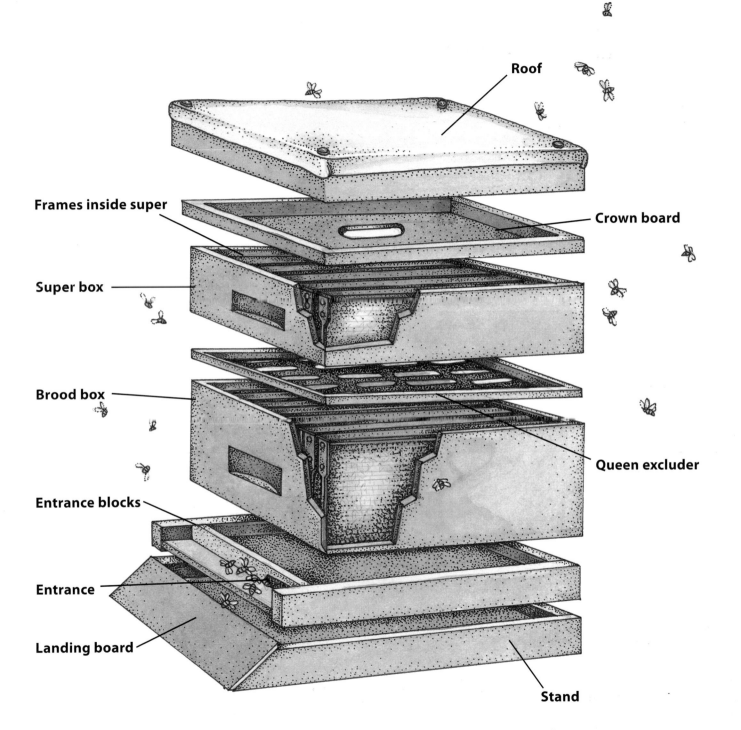

Roof

Frames inside super

Crown board

Super box

Brood box

Queen excluder

Entrance blocks

Entrance

Landing board

Stand

Different types of hives

The double-walled William Broughton Carr (WBC) Hive

Ask most people to draw a picture of a beehive and they will draw what we call a WBC hive (pictured). William Broughton Carr was a keen amateur apiarist at the turn of the 20th century and designed a hive to suit the climate of the British Isles. With its gabled roof and sloping sides, it conjures up images of apple blossom, buttercups, gently flowing rivers and willow trees swaying in the breeze. Who wouldn't want one in their garden?

A professional beekeeper, or even a keen amateur, will tell you that the WBC hive is expensive and unwieldy. However, it is the most attractive hive, so you will have to weigh aesthetic considerations against practicalities. Does adding such an attractive feature to your garden outweigh the disadvantage of any extra work involved?

Some of that extra work comes from the fact that the WBC is double walled. The outer parts (lifts), which we see as the sloping sides, act principally as insulation and as a casing for the inner parts. They have to be completely removed before you can get to the inside brood boxes and the supers. This does involve a bit more time and effort. On the other hand, since the brood and supers of a WBC are protected by the lifts, they are built with thinner wood than other, exposed hives. This reduces the weight of the inner boxes, which can be seen as an advantage.

The floor which both structures sit on acts as the stand and entrance with an attached landing board. The floor can also accommodate a Varroa screen for use in an integrated pest-management scheme (see page 121).

The bees' entrance to the WBC hive has a porch which keeps off the rain and accommodates the entrance blocks. You can easily slide these blocks open or closed to change the size of the entrance – you may want to make it smaller in the winter to stop mice getting in, or larger when the weather is hot to help ventilation.

The attractive gabled roof completes the hive. It contains ventilation holes and, since it overhangs the structure, it gives the hive that

Above *The Commercial hive and right, the Langstroth. Although they may look similar, they have their own bespoke parts that aren't interchangeable.*

Single-walled box hives

extra bit of protection against the elements by keeping rainwater off the lifts.

Like other hives the WBC can be built from most wood materials, but cedarwood is recommended. Cedar's built-in oils and preservatives make it a great wood for the outdoors, as it can last decades even in extreme climates. It used to be the fashion to paint the outside of the hives white, but nowadays it is common to see the cedarwood left in its natural state. The bees don't mind either way.

Single-walled box hives are by far the most common type used worldwide by both professionals and hobbyists, and the components are easily bought from beekeeping suppliers. There are a number of designs and each region has its favourite. In some parts of the world there will be little choice on offer; in others you will find that there is a selection of four or five. The important thing to remember is that the component parts of different hives are unique to that design, so if you are buying extra frames or boxes, make sure that they are compatible with what you have already.

Box hives aren't as attractive as the WBC – in fact they look like a stack of boxes – but they are easier to use and the frames have more surface area, which accommodates more bees. For this reason, new beekeepers may choose to forgo the aesthetics and start with a single-walled box hive.

In practical terms, the style of hive you start with is likely to be the style you continue to use since you can't interchange the parts of different hives.

Box hives all work on the same principle – stackable floorless boxes which hold the frames – and look very similar, but they vary slightly in size. The single-walled version is made with thicker wood parts than the WBC: since the box is open to the elements, the wood needs to be thick enough to offer some degree of insulation from both the cold and the heat. The roofs and floors are separate items and the hives require a stand to keep them off the ground.

There are at least six different commonly used styles of box hives. This relatively high number comes down to eager enthusiasts with the time and money to devise new designs, and commercial chaps intent on making the most efficient honey-producing hives. Those who were successful in marketing their modifications gained a degree of following, which has led to some national preferences. The Langstroth is the most commonly used hive in the USA, the National in the UK and the Dadant in France. That is not to say that the Commercial, WBC and Smith are not used, but they are less popular.

The Dadant and Commercial are the largest and heaviest of these hives and, as the name of the latter suggests, are usually left to the professional apiarist. Our advice to the hobbyist is to use the lighter, more manageable hives.

Remember the 'bee space' (see page 30)? Well, some hives are 'top' bee space and some are 'bottom' bee space. The National, for example, is a bottom bee space while the Langstroth is a top. This means the space for the bees to crawl through to get around the hive is either below the frames or above the frames. This makes no practical difference to the workings of the hive, but it does highlight the fact that the various hive types are not interchangeable.

Above *The UK's most popular hive – the National.*

Other hives

Top bar hives (above) are used in many parts of Africa, where they can be built using cheap local materials. They have a movable frame but rather than using a constructed frame with foundation, the bees are encouraged to build their own honeycomb on bars, which are removed completely when the time comes to harvest the wax and honey.

A variation of the top bar hive is the long deep hive. Using the same frames as the single-walled hives, this design has been developed with the beekeeper – rather than the bees – in mind. Instead of growing vertically, the hive expands horizontally. It can be mounted higher off the ground so the beekeeper doesn't need to bend over so much and is designed to minimize the lifting of heavy supers.

tip

Nowadays it is common to see the cedarwood left in its natural state. The bees don't mind what colour it is!

How to make a frame with foundation

The principle for putting together a frame is the same for the brood and the super frames, since the only difference is their depth. The frame consists of a top bar, two side bars, two bottom bars and the nails to hold them all together.

The top bar has a holding bar that needs to be removed before frame assembly. It will be attached later to hold in the foundation. You can use the hive tool to remove this holding bar, but be careful – the sharp edge can easily cut a finger.

Once you have removed the holding bar, attach the two side bars to the top bar (this is a tight fit) with the grooves facing inwards. Place a bottom bar into the side bars (on the same side of the frame as the thicker side of the top bar). Square the frame up and nail these pieces together.

The frame is now ready for foundation. If you are using wired foundation, bend the three top loops to 90 degrees and the two bottom loops back on to themselves so that they won't protrude from the bottom of the finished frame.

Starting with the edge which has the three loops, insert the foundation into the grooves of the side bars, making sure it stays flat, right.

Once the three hoops are touching the top bar, replace the wedge bar to hold the foundation and hammer three nails diagonally through the holding bar and the wire loops (left).

For unwired foundation just insert the foundation into the frames and hold it in place with the wedge bar and three nails.

The bottom of the foundation should sit on the bottom bar but not protrude out of it.

Fit the second bottom bar into the side bars and nail from the bottom. Ensure that the frames are straight.

> **tip**
> The principle for putting together a frame is the same for the brood and the super frames.

Right and above *Putting together a frame is a simple task, and is the same principle for the brood and super frames – the only difference being their depth.*

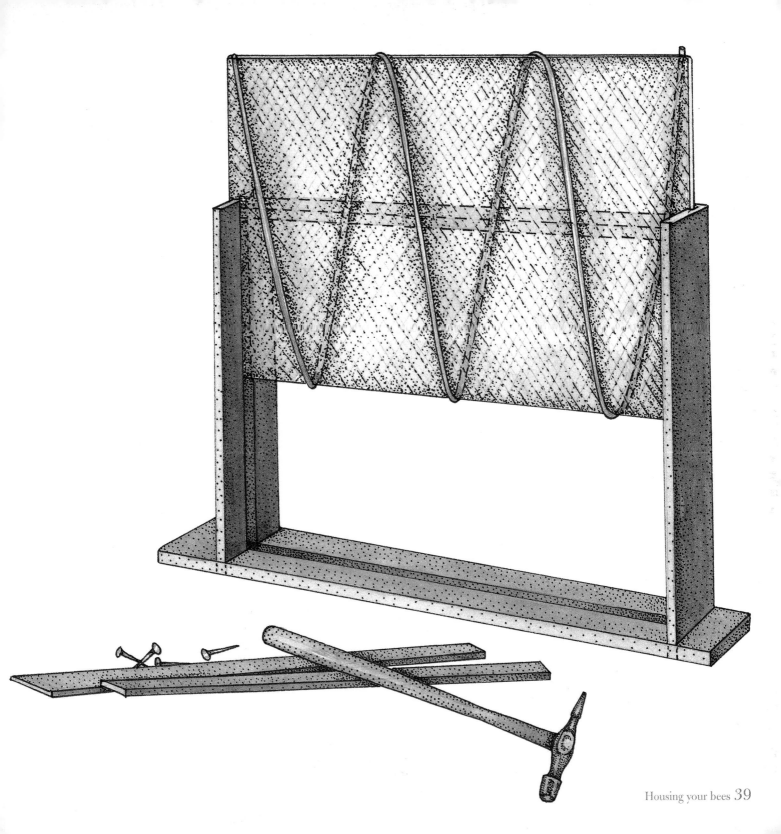

What you will need

You will enjoy beekeeping when you feel comfortable and safe around your bees. For this, protective clothing is a must, along with certain pieces of equipment including a smoker and a hive tool. There is a wide range of protective clothing available, from all-in-one suits to smocks and veils. The suits are more expensive but do offer better protection. As well as traditional white cotton, they now come in lighter materials and other colours. Some manufacturers even offer a made-to-measure service. You will also need to think about honey-harvesting equipment. Manual honey extractors are adequate for the hobbyist with a few hives. An electric extractor is a pricier option.

The Kit

Protective clothing

Looking after your bees will be stress free and relaxing when you feel safe from the possibility of being stung. There is a variety of protective clothing available from beekeeping suppliers but whatever you choose to wear the idea is to present yourself to the bees in the least threatening way and to keep your skin covered.

The best way to achieve this is to kit yourself out with a full body suit. These all-in-one suits may make you look like a spaceman, but they do provide total protection, leaving you confident to get on with the work in hand. The basic type comes in white and is made of cotton, zipped at the front and with a veiled hood attached.

Alternatively, you can buy a suit made primarily from lightweight olive-coloured polyester. Its veiled hood can be unzipped at the front so that when you lift it over your head it lies comfortably at the back. (With some other suits you have to take the hood off completely, or else it dangles around your chest, getting in the way.) This design is not the cheapest on the market, but is by far the most comfortable, the least conspicuous and, importantly, the coolest. It can get hot out there inspecting your hive in the middle of a summer's day. Suits come in different sizes for adults and children, and some manufacturers offer a tailoring service at no extra cost.

The price for all-in-one suits varies between manufacturers, so do shop around. They are certainly worth the investment. A cheaper option is to buy a combined hat and veil These incorporate a hat, netting and neck and shoulder coverings. The material can be tucked into a shirt or jumper, giving a great degree of protection without inhibiting your vision. In addition you can buy jackets or smocks that can be tucked into trousers, which are also sold separately.

The reason bee wear is light-coloured and smooth in texture is that bees don't like bears – big, dark, furry animals who steal their honey. So it is unwise to wear dark, woolly jumpers around bees, as this awakens their primeval instincts. Bees also like to climb upwards into dark spaces, so leaving your ankles and cuffs open and loose is inviting one to come and explore. It could then find itself trapped, stressed, under pressure and near to your flesh. This is likely to leave you with a barbed sting pumping a dose of histamine into your body, which is best avoided. Don't forget to cover your ankles with socks; stings on the ankle are memorable!

Another part of the body that needs protection is the hands. The best bee gloves are made from a good-quality goatskin. Although they offer complete protection, with long wrist coverings and thin leather, they are hard to clean and you do lose finesse. There are many beekeeping situations in which you will need to handle things gently or there are fiddly manoeuvres to accomplish. Common household rubber washing-up gloves protect the hands well while allowing you to maintain a delicate touch.

Some very experienced beekeepers are happy using their bare hands. If the bees are calm and your movements are gentle, you may get away without being stung.

Right *Bee suits come in lightweight olive-coloured polyester, as well as the more traditional, heavier, cotton.*

tip
A full bodysuit may make you look like a spaceman, but it does provide total protection.

The smoker

The image of a traditional beekeeper is of a bearded man wearing a veil and holding a strange contraption called a smoker. While the beard may be in retreat as more women and younger people take up the hobby, the Victorian-looking smoker is still an important piece of equipment. To understand why you use a smoker, you need to get inside the brains of your bees.

Bees can protect themselves against living creatures by using their sting and venom, but since a common natural habitat of a honey bee is a hollowed out tree there is another threat that presents a great danger and against which it can't use its sting – fire.

Bees are alerted to fire by the smoke that precedes the flames. Their instinct is to vacate the hive, get to a safe place and start a new home. But before they leave they spend as much time as they can collecting their worldly possessions, which in their case is the precious honey that they have been so busy storing. Diving into the comb, they quickly suck up

HOW TO USE A SMOKER

The smoker is designed to produce cold smoke. It is not meant to erupt and spew out flames or heat. The bees just need to get a hint of smoke to believe there is danger about. The fuel that is placed inside the smoker is meant to smoulder rather than burn. This is why its canister has only a small hole at the bottom, which restricts the amount of air getting into it. The bellows are there to boost the airflow, which will increase the heat and set light to a bit more of the combustible material inside, allowing it to smoulder more and producing the cool smoke you desire.

So what type of material can you use that will produce lots of smoke, without too much heat, and will continue to smoulder and not go out just at the moment you need to calm the bees down?

Hay, grass cuttings and wood shavings all do the job. Cardboard works very well when rolled up into a cylinder with the veins running vertically, creating an airflow. Combining cardboard with hessian sacking – the type that is used for sand bags and vegetables – makes an excellent fuel that can be made up in advance. Hessian is not as common as

it used to be before the advent of synthetic materials, but it is still available from beekeeping suppliers. Cut both hessian and cardboard to the height of the smoker and roll them up together, a layer of hessian wrapped in a layer of cardboard. Light the material, push it into the smoker, pump the bellows a couple of times and you should have a working smoker that will last 30 minutes or so.

As with any material you propose to use in a smoker, ensure that it has not been treated with fire-retardant chemicals, as is sometimes the case with cardboard packaging. The bees will not take kindly to the noxious smoke produced by this type of fuel.

A gentle puff into the entrance five minutes before you open the hive should give the bees enough time to get them busy sucking honey. They will generally move away from smoke, so the smoker can also be used to herd them off a frame. Too much smoke, however, can make them angry and likely to sting.

Using your smoker effectively is an important aspect of the art of beekeeping. It can be a bit tricky at first, so it is worth spending time practising until you get it right.

Where and when to get your bees

Once you have set up and located the hive, bought your suit and equipment, and have a basic understanding of how bees live and work, you will need to find a supplier of good-quality bees. Many suppliers and commercial farmers advertise in beekeeping magazines, but where possible go for word-of mouth recommendations. Alternatively, your beekeepers' association may be able to provide you with a swarm. In either case, early summer is the best time for new bees. You will need to know in advance how many bees you are getting and what type they are.

Making your bees feel at home

The bees in a newly purchased nucleus are likely to be tired and hungry from their journey. A 'welcome meal' will give them the energy to set up their new home, while watching and listening to them coming and going can tell you a lot about their wellbeing. The foragers will seek out the best sources of nectar and pollen and if their back legs are covered in pollen you know that they are busy bringing in food for the hive. Soon the colony will outgrow its living space. Inspect it after nine days and every seven days thereafter, to judge when to add an upstairs extension. As long as your bees have enough room during the summer, you don't disturb them too often and the weather is warm, they should build a strong and healthy colony.

Pollen

Honey bees and the flowers they visit are a great example of the interconnected relationships that occur in nature. Plants need bees to pollinate them and bees need plants for food. If this connection were lost, they would both disappear. Thankfully the bee is very good at her job and there are plenty of plants for her to visit. Each plant species has a different nectar which gives each crop of honey its differing flavour, colour and consistency (see page 80).

One of the most fascinating sights for any novice beekeeper is the forager bees returning to the hive, their back legs bearing a colourful load of pollen. It looks far too heavy for them to carry, but is actually much lighter than a stomach full of nectar. Although bees are only one of many insects that are responsible for pollination – others include butterflies, moths and even wasps – they are the only ones whose bodies are specially adapted to collect pollen, and the only insects who feed their young pollen, which is rich in the protein, fats and lipids needed for the bees' development.

Pollen is the male gamete of the plant and, in order to reproduce, it needs to mate with a female gamete of the same species. The bee transports the pollen grain from the male to the female by collecting minute particles on the hairs of her body as she visits the flowers, collecting the nectar or the pollen.

It is one of the mysteries of bee communication that a forager bee knows the purpose of her search before she leaves the hive. If she is going out to collect pollen she will take enough honey in her stomach to act as fuel for her flight, plus a bit more to use as glue to bind the pollen together. If she is going out for

nectar she can stay out longer, since she will collect the energy-giving liquid on route.

As the honey bee is a methodical pollinator, she generally forages on one type of flower until she has exhausted the supply – unlike butterflies, which can flit from one flower type to another in the course of one flight. Both pollen loads on a bee's legs will therefore be the same colour and from the same source. Back at the hive this doesn't matter, since the pollen is mixed together as food for the bees, but the plants benefit, as this approach gives them the greatest chance of being pollinated.

It is fun to try to guess, with the help of a handy pocket pollen-colour guide, which flower a bee has visited, according to the colour of the pollen and the time of year. Each area has its own pollen cycle, running from early spring through to late autumn. In Britain,

Above *A bee with back legs laden with pollen.*

Right *Fruit growers will hire local bee hives for a few weeks each year to pollinate the crop. Bees will naturally go to the nearest and most bountiful source of nectar, so pollination is easy to promote, as is honey from a single type of flower such as orange blossom, lavender and so forth.*

it tends to start with crocuses, followed by dandelions, fruit blossom trees and horse chestnut, then lime trees, clovers and late-summer plants, before autumn garden flowers such as Michaelmas daisies, and finally ivy. While most pollen is a shade of yellow or orange, ranging from the very pale to the very bright, some is surprisingly brick red (horse chestnut), black or deep purple (oriental poppy) or light green (hawthorn).

A large amount of one particular kind of pollen, however, does not mean that honey will be made from the nectar of that plant. Some plants are a much richer source of pollen than of nectar and vice versa.

When the bee returns to the hive she passes her load to a waiting house bee, who inspects a number of cells before dropping the load into a suitable one. This is usually located just above or beside the brood, so that the pollen is readily available to feed the older larvae, the nurse (newly born) bees and the younger hive bees. House bees will add some honey to the dry pollen and seal the cell with some more. This mixture is called bee bread and it preserves the pollen for later use.

A colony's intake of pollen can fluctuate according to the weather and the availability of pollen-rich plants. To prevent demand outstripping supply during a bad spell of weather, a colony will stockpile around a week's worth of pollen, some 1kg (2¼lb).

Commercial pollination

As honey bees can be moved from one place to another, they play a vital role for commercial fruit growers and farmers, and are valued more highly as pollinators than as honey producers. In the USA, for example, where honey bees are moved huge distances to pollinate vast stretches of almond trees in California, apple trees in Pennsylvania and blueberries in Massachusetts, their value to the economy is measured in tens of billions of dollars. In the UK, fruit growers in Kent, jam manufacturers in Cambridge and West Country cider producers will often have their own hives or bring in local apiarists for pollination. As a result, the bees' value to the UK economy is close to £1bn. Lack of bee pollination can result in misshapen fruits and 30 per cent lower yields of many fruits and vegetables. If you have a vegetable patch, you will be pleased to know that bees are particularly good for pollinating marrows, pumpkins, cucumbers and tomatoes.

tip
Each area has its own pollen cycle, running from early spring through to late autumn.

POLLEN FACTS

- **32kg (71lb) pollen has been recorded from a single colony.**
- **To rear one worker bee requires ten average bee loads, the equivalent of 0.14g (¹⁄₂₀₀oz) of pollen.**
- **One average bee load weighs 0.014g (¹⁄₂,₀₀₀oz).**
- **450g (1lb) of pollen rears 3,250 bees. This means that 27kg (61lb) of pollen is required to sustain an average colony throughout the year.**
- **This equates to some two million pollen-collecting trips.**

Gardening for bees

Nothing evokes summer quite as much as the buzz of bees collecting pollen. Unfortunately, some of our most popular cultivated flowers provide no nectar and little pollen. Encouraging honey bees to stay in the garden or attracting other types of bees may require some horticultural changes. Culinary herbs and fruit trees are bee-friendly, and many wildflowers often regarded as troublesome weeds provide a rich source of food. Leaving a patch of your garden to run wild can bring much pleasure to your bees and be a welcome habitat for other creatures. The most useful flowers you can plant are those that provide food early and late in the year. Not only will they supply nectar and pollen to your bees when these are in short supply, but they will also add colour to your garden.

Some suggested flowers for a bee garden

Annuals

Baby blue-eyes (*Nemophilia spp*), bindweed (*Convolvulus spp*), blanket flower (*Gaillardia spp*), blue-eyed Mary (*Collinsia violacea*), California poppy (*Eschscholzia californica*), candytuft (*Iberis amara*), China aster(*Callistephus chinensis*), clarkia (*Clarkia elegans*), common baby's-breath (*Gypsophila paniculata*), cornflower (*Centaurea cyanus*), flax (*Linum* spp.), gilia (*Gilia* spp.), godetia (*Clarkia amoena*), lavatera (*Lavatera* spp.), love in the mist (*Nigella* spp.), mallow (*Malva* spp.), malope (*Malope trifida*), meadowfoam or poached egg plant (*Limnanthes* spp.), Mexican aster (*Cosmos bipinnatus*), mignonette (*Reseda* spp.), Nasturtium (*Tropaeolum* spp.), phacelia (*Phacelia* spp.), pheasant's eye (*Adonis* spp.), sweet alyssum (*Lobularia maritime*), tickseed (*Coreopsis* spp.)

Perennials

Alpine rock-cress (*Arabis alpina*), alyssum (*Alyssum* spp.), aubrieta (*Aubrieta* spp.), Canterbury bells (*Campanula medium*), Carpathian harebell (*Campanula carpatica*), catmint (*Nepeta* spp.), checkerbloom (*Sidalcea* spp.), coneflower (*Rudbeckia* spp.), cranesbill (*Geranium pratense*), fernleaf yarrow (*Achillea filipendulina*), fleabane (*Erigeron* spp.), forget-me-nots (*Myosotis* spp.), French honeysuckle (*Hedysarum coronarium*), fuchsia (*Fuchsia* spp.), golden rod (*Solidago* spp.), herb hyssop (*Hyssopus officinalis*), hollyhock (*Alcea rosea*), horehound (*Marrubium vulgare*), hound's tongue (*Cynoglossum* spp.), knapweed (*Centaurea* spp.), large blue alkanet (*Anchusa azurea*), lavatera (*Lavatera* spp.), lavenders (*Lavandula* spp.), loosestrife (*Lythrum* spp.), marjoram (*Origanum majorana*), sage (*Salvia* spp., especially *S*. x *superba*), scabious (*Scabiosa* spp.), sneezewood (*Helenium* spp.), thoroughwort (*Eupatorium* spp.), thrift (*Armeria maritima*), thyme (*Thymus* spp.), veronica (*Veronica* spp.), willowherb (*Epilobium* spp.), woundwort (*Stachys* spp.)

Bulbs

Autumn crocus (*Colchicum autumnale*), common hyacinth (*Hyacinthus orientalis*), crown imperial (*Fritillaria imperialis*), quamash (*Camassia* spp.), snowdrop (*Galanthus nivalis*), snowflake (*Leucojum* spp.), tulip (*Tulipa* spp.), winter aconite (*Eranthis hyemalis*

tip
The most useful flowers you can plant are those that provide food early and late in the year.

Above *It is most beneficial for the bees if you plant a variety of flowers that between them will provide year-round food.*

All about honey

It is probably the prospect of home-grown honey that entices most people to keep bees. The sweet golden liquid that has enchanted mankind for centuries is one of nature's glorious gifts. What could make a more personal present that honey from your own hive? One of the most fun times for all the family is the honey harvest, when the oozing substance is ready to eat. When you will get most of your honey, how much you will get, and what it will taste of are often difficult to predict. Your bees will constantly surprise you. How you harvest and jar your honey will depend on whether you intend to keep it, sell it or show it. Whatever you decide, the right preparation will make all the difference.

Throughout the summer, the bees collect enough nectar to store and feed the colony during the cold, nectar-drought months of winter. It is these stores that you are in effect stealing from the hive when you harvest the honey.

A strong colony of honey bees gets busy collecting nectar as soon as the weather is warm enough for the plants to start producing it. When the fruit trees have blossomed and the dandelions are out you can be sure the foragers will be out and about too. They may travel up to 5km (3 miles) from the hive in their search, but are thought to favour the nearest source. When a scouting foraging bee finds a good source of nectar, she uses the waggle dance to let the other foragers know (see box). Bees are not fortune-tellers and the vagaries of the weather can produce periods where collection is not possible – a rainy spell or even a heat wave, when flowers conserve their nectar – so they collect as much as they can, when they can. The amount depends on the availability of the nectar and how much storage space is in the hive.

As well as being used to feed the larvae, honey gives the foragers fuel for their nectar- and pollen-collecting flights, and the house bees energy to keep the brood areas of the hive warm. To keep the maternity wing and nursery at a constant temperature of 35°C (95°F) the bees huddle together and shiver. This movement of the body, which increases the temperature to the required heat, takes a lot of energy. If the outside temperature is too high, the bees use their wings to fan air throughout the hive to keep it cool. Again this uses a lot of energy and they get this by eating the honey they have been so busy collecting and storing.

Bees also need to eat honey to stimulate the production of wax.

During a year a colony requires at least 120kg (265lb) of honey just to be able to function.

> **tip**
> Bees are not fortune-tellers and the vagaries of the weather can produce periods where collection is not possible.

THE WAGGLE DANCE

The returning forager bee is laden with her bounty and eager to tell her sisters where the supply can be found. First she climbs on to the vertical comb and lets the other bees taste a bit of the treasure – this could be nectar, water, pollen or propolis. Once they have had a good sniff to assess the quality and know what they are going to be looking for, they stand around watching, listening and smelling the bounty hunter as she does her waggle dance.

This shaking of the body as she repeats a 'figure of eight' pattern tells the onlookers the direction and distance of the new supply. If the bee waggles her body for a fraction of a second the source is near, but if the waggle lasts for a couple of seconds the source will be around 2,000m (1 1/4 miles) away. The direction of the source is conveyed by the direction of the dance in relation to the sun's position. Think of a clock face on the honeycomb. If the dancer's head points upwards towards 12 o'clock, she is telling the other foragers that the source lies on a line directly towards the sun. When, after an hour, the sun has moved 15 degrees across the sky, the forager will need to alter her dance direction by 15 degrees. So this time the dance will not be straight up on the honeycomb; instead, the head will be pointing halfway between 11 and 12 o'clock. When the new foragers fly from the hive they first get a sense of where the sun is and then fly in the direction told to them by the waggle dancer – in this case 15 degrees to the left of the sun. If at another time the source of food was opposite the sun, the dancer would waggle with her head pointing downwards.

The waggle dance isn't totally accurate, but enough to get the bees near the new source. When they get there they may have to hunt around for a few minutes before they recognize the food by its smell.

It is a remarkable form of communication for any animal, let alone one with a brain the size of a breadcrumb. We mustn't forget to thank Karl von Frisch, who won a Nobel Prize in 1973 for his work on this subject.

The waggle dance isn't totally accurate, but enough to get the bees near the new source.

How much honey?

During the course of the summer the colony will have been expanding as the queen lays thousands of eggs to increase the army of foragers who bring back the food to feed the hive. In the wild, the honey the bees produce at the beginning of the season is used by the emerging population, which increases five-fold over the summer. Towards the latter half of the year the queen reduces her laying and the colony prepares for winter, getting the stores in to keep them going over the cold months.

Things are a bit different for a colony that is kept in a hive and managed by a beekeeper. Since you can keep adding extensions to the hive, the bees find themselves with more room to fill. It is their instinct to fill available space with honey stores and the more space available the more eggs the queen will be encouraged to lay. So the amount of honey you get from a colony depends in part on how you manage the hive. Other factors that play a role include the weather, the availability of good forage, competition from other hives in the vicinity, any ailments the bees may have and the character of the bees themselves.

It is estimated that you can expect anything between 9 and 18kg (20 and 40lb) of honey from one hive in your first year of beekeeping. Just one national super with ten frames of capped honey can weigh around 10kg (22lb). So even if you think you have bought plenty of jars, don't be surprised if you end up having to borrow containers of all shapes and sizes from family and friends.

Above *Cities are awash with nectar for the bees to feast on. With a huge diversity of flowers and plants to visit, urban honey is often more flavoursome than other varieties.*

Right *Oil-seed rape covers huge swathes of land with its bright yellow flowers which bees love, quickly making honey from it in late Spring.*

When you get the honey

When your honey is ready to harvest depends primarily on the types and abundance of flowers, crops and trees within a 5km (3 mile) radius of your hives. But so many other factors come into play – most importantly the weather – that it is impossible to promise anyone a jar of home-grown honey by a certain date. If the colony is strong and the conditions have been good, you may find that a super or two will be full by midsummer. You may also get a late harvest towards the end of the summer. Each hive will perform differently.

Even two hives standing next to each other can produce different amounts of honey at different times.

It is also hard to predict what your honey will taste of unless the hives are located near fields of a particular source of nectar. If, for example, there is oil-seed rape nearby, there is little chance of your honey tasting of anything else. Bees love this bright yellow flower which is increasingly taking over vast swathes of the countryside. Its nectar is available in the spring and they can work it very quickly, so you may get a super's worth of oil-seed rape honey by

late spring. However, the downside of oil-seed rape is that it crystallizes quickly and the bees then find it difficult to eat. Remove it within a few days and extract it (see page 84) before it crystallizes on the frame.

The predominance of oil-seed rape in some parts of the countryside means that many city honeys are more flavoursome and delicate. Urban bees, having enjoyed the biodiversity of plants and flowers that gardens, parks and railway cuttings have to offer, can often produce honey with a much richer texture and taste than their rural sisters. Lime trees line many city streets, so it is no surprise that late-summer honey from an urban hive often has a slight citrus taste. A bumper harvest of elderflower honey in early summer after a very wet spring is more unusual. Bees turn to elderflower only in desperation, but the honey is delicious with a light, fragrant bouquet – the one consolation for all the rain!

Not knowing what your honey will taste of from one harvest to the next, let alone one year to the next, makes the honey harvest really exciting as you try to guess what plants your bees have been feeding on and how the honey will compare to previous batches. The downside is that you can't promise people more of, say, the early summer crop they particularly liked the previous year, because you can't guarantee that this year's honey will taste anything like it.

Unlike wine tasting, there don't seem to be courses in honey-tasting around, but you can still enjoy the challenge of trying to decipher the nuances and subtle flavours of your polyflowered honey.

The taste of honey

Honey's subtle taste and flavour is mainly determined by the plant from which the nectar is obtained. This also shapes the aroma, density and colour of the honey. In many countries, honey comes from a single plant such as clover in Canada and New Zealand, eucalyptus in Australia, leatherwood in Tasmania, lavender and rosemary in France or Spain, wild thyme in Greece, buckwheat in the eastern US and Russia, orange blossom in Florida or California, and acacia in eastern Europe. In Britain, most honey – with the exception of heather and some white clover honey – is usually a blend of wild flowers, including hawthorn, fruit blossom and the clover-like sainfoin. However, in urban areas, where trees in parks and lining roads are the predominate source of nectar, slight citrus-flavoured honey from the lime and the sycamore tree are common. There is no direct correlation between the flavour of a honey and its colour, but the lighter the honey, generally the milder in flavour it is than darker-coloured varieties.

Plant	Colour
Acacia tree	Very light yellow, almost transparent
Alfalfa	Amber to biege
Apple blossom	Light amber
Balsam	Amber
Borage +	Pale yellow
Buckwheat	Purplish black
Chestnut	Very dark brown
Clover	Varies from white to light amber
Dandelion #	Deep yellow
Elderflower -	Very pale yellow
Eucalyptus	Light amber
Fir tree	Dark brown
Lavender	Dark amber
Leatherwood	Amber
Lime blossom *	Very light amber with greenish tinge
Hawthorn	Dark brownish
Heather	Reddish brown
Orange blossom	Light amber
Oil-seed rape #	Light yellow
Raspberry	White with a red hue
Rosemary	Light amber
Sainfoin	Very light, almost white
Sycamore tree *	Amber with greenish tinge
Sunflower #	Opaque, egg-yolk yellow
Thistle	Deep yellow with green hue
Wild thyme	Amber
Willowherb x	Pale, water-white

+ Also known as star flower honey as borage has sky-blue star-shaped flowers
# Honey crystalizes very quickly so has to be harvested early
- Very unusual but in a particularly bad Spring when no other nectar sources around the bees will visit the elderflower
* One undesirable feature of the lime tree and the sycamore is honeydew, which is the secretion of insects that live on the leaves. On occasions the bees collect the honeydrew, but in many European countries, especially Germany, the most highly sought after honey is honeydew from conifers.
x Also known as fireweed

Taste	Consistency	Time of year for nectar
Very mild, hint of vanilla	Thin	Mid summer
Delicate and fruity	Thick	Late summer
Hint of apple	Thick	Early summer
Toffee flavoured	Thick	Late summer
Mild, hint of camomile	Thin	Mid summer
Strong, malty	Very thick	Early summer
Treacle and molasses, bitter, rich in pollen grains	Very thick	Late summer
Creamy butterscotch	Medium	Mid summer
Hint of ammonia	Thick	Spring
Elderflower cordial	Thin	Spring
Hint of toffee and raisin	Thick	Summer
Malty, toffee-like, with hint of aniseed	Thick	Summer
Strong 'flowery' perfume	Medium	Mid/late summer
Malty, spicy, woody aftertaste	Medium	Mid summer
Citrus/minty	Very thin	Mid summer
Rich, nutty, almondy	Very thick	Early summer
Hint of toffee	Thick, jelly-like	Late summer
Hint of citrus, tangy and very sweet	Medium	Late spring
Bland	Medium	Late spring
Raspberry	Unctuous	Mid summer
Strong aroma, hint of rosemary	Medium	Early summer
Slight fruity taste	Thin	Mid summer
Hint of citrus	Fair	Mid summer
Suggestive of lemon	Thick	Late summer
Slightly bitter and astringent	Medium	Late summer
Intensely aromatic, spicy, touch of thymol	Medium	Late summer
Bland	Medium	Late summer

How to harvest honey

It has been said that the enjoyment of beekeeping can be spoiled by the hassle of harvesting honey, but with some planning and the right equipment the process can be quick, straightforward and above all lots of fun – as long as you accept the fact that not only your equipment but you too are going to get sticky. Fortunately honey is easy to clean off with warm soapy water. Wax is a bit more troublesome, so have plenty of newspaper or old tablecloths around to put on your worktops and floor, and keep dedicated bowls and utensils for the job.

Ideally you will have a utility room with a sink and lots of space in which to harvest your honey, but we don't all have that luxury. You may well have to work in a small kitchen, in which case keep the doors and windows closed to keep out any unwelcome intruders, particularly inquisitive wasps. Try to avoid harvesting your honey outside, as you will soon be surrounded by bees and wasps.

The plan is to get the honey out of the frames into containers without bits of wax or bees in it, so you need to work in a systematic way: get your frames full of capped honey into the house without the bees, open up the cappings, extract the honey, strain it to get rid of the unwanted bits, then let it settle before pouring it into jars. Remember that full supers can be very heavy – around 10kg (22lb) – so if you are a lone beekeeper you may need some assistance carrying them.

Getting the frames without the bees

The first step is to get the bees out of the super that you want to harvest. You really don't want lots of bees clinging on as you try to remove the honey from the frames.

To do this, you can use a clever little device called a bee escape. There is a variety of types on the market but they all work on the same principle – they let the bees leave the super but don't let them return. In other words, they have a one-way valve.

Place the bee escape in the hole of the crown board and place the crown board under the super you want to remove. After a few hours the bees will have moved downstairs and been unable to get back up, so the super should be bee-free.

Comb honey

Once you have got your super safely inside, it is time to decide if you would like to keep some comb honey, or if you want to extract all the honey from the comb for liquid honey. Most people like a bit of both.

For comb honey, simply use a thin, sharp knife and slice small chunks, perhaps 10 x 7.5cm (4 x 3in) of the comb straight from the freshly collected frames. If you have been using wired foundation, cut the wire and pull it out of the comb before putting the comb into your containers. If you think that you may want more comb honey in the future, you will need to buy unwired foundation. The wires are only there to give strength when the honey is spun in an extractor (see below). You may like to put some comb in a jar and fill it up with liquid honey. This is called chunk honey and it makes a great present for people who have expressed a preference for honey on the comb.

Scraping off the honey

If you don't have access to an extractor, you can harvest your honey by scraping the comb. It takes longer and is messier than using an extractor, and it uses more utensils, but has the slight advantage that less air gets mixed in with the honey. The main disadvantage is that it destroys the honeycomb. Ideally you want to give this back to the bees, as it cuts down their workload: instead of having to draw out new foundation, they are able to store nectar immediately in the honeycomb they made during the year.

If you are going to use the scraping method, you will need a large pot to collect

Above *Empty honeycomb after the honey has been extracted.*

the honey, a large container for the comb, a large spoon or scraper and a coarse sieve. Later you will need a finer sieve to filter the honey a second time and another large pot or bucket – preferably with a valve (honey-gate) at the bottom so your honey can easily flow into jars. These honey buckets or tanks can be bought from many beekeeping suppliers. Place the coarse sieve over the large pot and, holding the frame of honeycomb over the sieve, scrape the honey straight off the frame, comb and all, into it. You are not going to be able to use the foundation again, so just scrape away as hard as you like.

The sieve will soon be filled with a lovely gooey pile of comb and honey. Let the honey ooze through the sieve into the holding pot. Once most of it has passed through, empty the wax comb scraping into the large container. (You can sieve this again later and use the remaining wax for making candles or lotions – see page 106.) Continue this process until all the frames have been emptied.

To get really clean honey, filter it through either a finer sieve or some muslin into another large holding pot or bucket. This will weed out the small particles of wax, bee bits and pollen that got through the coarse filter.

After the second sieving process is complete, put a lid on the container and let the honey stand so that the air bubbles can rise to the top. This may take as much as a couple of

tip
You can sieve the wax comb scraping again later and use the remaining wax for making candles or lotions.

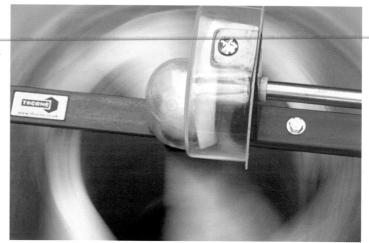

days, depending on the consistency of the honey and the ambient temperature.

You now have empty scraped frames that are wet with honey. The best thing you can do with these is to return them to the bees for cleaning. Put them back in the super and place the super above the crown board (as long as the hole is open). In a day or two the frames will have been licked clean and you can use them again to put new foundation in.

Using an extractor

An extractor makes the process much easier, quicker and less messy. It can be costly to buy, so it is worth asking your local beekeepers' association if they have a communal extractor for hire. Many do, but there may be a waiting list at weekends at busy times of year.

There are a variety of extractors on the market. The cheaper ones are manual and plastic. More expensive versions have an electric motor and are stainless steel. In either case, extractors are either tangential or radial. In a tangential extractor you have to turn the frames in order to empty the honey from both sides. Radial extractors do both sides at once. If you have no more than a handful of hives, a manual tangential extractor will suffice. Whichever one you decide to use the principle is the same. A number of frames of uncapped comb are placed in the device and spun either by you or by the motor. The honey 'flies' out of the comb on to the sides of the extractor, collecting at the bottom of the tank. The collected honey can then be filtered into another honey tank or bucket for settling and later bottling.

For this method of honey extraction you will need an uncapping device (either a knife or a fork), a honey tank/bucket – preferably with a honey-gate at the bottom, as this makes it easier to pour the honey into jars –

and a double strainer. The uncapping device opens the caps of the honeycomb, allowing the honey to spill out of the frames when spun in the extractor. It's up to you whether you choose a fork or a knife – try both and see which you are more comfortable with.

Run the fork or knife just under the surface of the cappings. You want to take off just the white cappings (left) and leave the rest of the comb intact. An uncapping knife works better when it is hot, so have some water on the boil to dip it into after each scraping.

Put the wax cappings into a container – you can melt them later to make candles (see page 108). When the cappings have been removed from each side of the frame, place the frame into the extractor.

Extractors normally hold at least four frames, so continue uncapping until the extractor is full. Put on the lid and turn the handle (or flick the switch) slowly at first. Let some of the honey spin out before increasing

the speed (so that the comb stays intact) and continue spinning for a couple of minutes. With a tangential extractor now is the time to rotate the frames to empty the honey from the other side, then carry on spinning for a few more minutes.

You will see that the honey is clinging to the side of the tank and collecting at the bottom along with bits and pieces of wax and pollen. When no more honey is coming off the frames and the spinning has become lighter, remove the comb and store it in the super (you can give it back to the bees to clean).

When all the frames have been spun, filter the honey through the double sieve into the honey tank or bucket (or any large container). It will take around a half an hour to sieve 9kg (20lb) of honey – about one full super's worth. Now cover and leave to settle for 24 hours or more to let the bubbles rise to the surface. Once they have been skimmed off your honey is ready to pour into jars.

Right When all of the frames have been spun, it is important to filter the honey through a double sieve into a honey tank or bucket. This will remove any bits in the honey, such as wax or bee parts. It will take around half an hour to sieve 9kg (20lbs) of honey.

How to jar honey

Choosing your jars

Your choice of jars depends largely on what you plan to do with the honey: are you going to eat it yourself or give some away as presents to family and friends; are you going to show it, or even try to sell it? If it's mainly for your own use, you may just want to store it in recycled jars that you have been keeping hold of throughout the year. A note of caution when using old jars, though: however much you wash them, lids often retain the smell of their previous contents and this will ruin the taste of the honey.

If you want friends and family to be able to share the enjoyment, you may want to buy a number of identical small attractive jars. Searching the internet should reveal suitable suppliers who will deliver to your door. Look for ones with tin-plate lids with a flowed-in seal (a built-in annular ring on the underside of the cap) which prevents air and moisture getting into the jars and ruining the honey.

Jars like this are also perfect if you want to sell your honey at a local delicatessen or farmers' market. If you want to show your honey, the type and size of jar will be stipulated in the entrance rules of the show (see right).

Putting your honey in jars

A mistake beginners often make in their excitement is not leaving the honey to settle for 24 hours or more in a honey bucket or tank. If you pour it into jars immediately after filtering, it will contain lots of bubbles. While these don't seem to affect the flavour, they do float to the top and leave an unsightly thin foam on the surface.

Once the honey has settled, pour it into the jars quickly – to reduce the amount of exposure to the air – but not so fast that air gets trapped. Getting this right becomes easier with practice and is a matter of judging how wide to open the honey-gate at the bottom of your honey bucket or tank. If your container does not have a valve, transfer the honey into a jug and then pour it from the jug into jars.

After you have filled a jar, put the lid on immediately. When they are all full store the jars a dark place, such as a cupboard or larder, at room temperature.

Labelling your honey

It's a good idea to label jars with the location of the hive, and the month and year the honey was made. If you are selling your honey or giving it away, include your contact details in case someone wants another jar. You can design your own labels on a home computer and print them out on sticky labels. It's not difficult to make them attractive as well as informative.

tip
It is vital that you leave the honey to settle for 24 hours after filtering, in a honey bucket or tank.

Left The honey-gate, or valve, at the bottom of the honey bucket has to be open wide enough for the liquid to pour quickly into the jars, but not so quickly that air gets trapped in the honey. This can take some practise to get right!

How to show honey

You may be curious to know just how good your honey is. Even though everyone who has tasted it says it looks delightful and it has a delicious flavour, how does it compare to other honey? Does it really have an exquisite taste or is it run-of-the mill? One way to satisfy your curiosity is to enter a honey show. Ask your local or national beekeepers' association or check the beekeeping journals. Most shows have classes open to beginners and children to encourage newcomers to have a go. Most charge an entrance fee and each class will have its own strict rules, so be sure to read them carefully and follow them to the letter.

Showing honey is not just about trying to win a ribbon, though that will of course be a tremendous accolade; it is also about meeting your peers, getting advice from the experts and learning how to improve your skills in honey production.

To be in with a chance of getting on the podium all aspects of your product must be top class, from the labels, which must conform to the show's guidelines, to the spotlessly clean jars, and the honey will need to be free of any unwanted materials such as bee parts or dust. Honey judges, especially at the top levels, will be able to tell if your honey has been spoiled by the soot of your smoker or scorched by an electric uncapping knife in the harvesting process. Therefore extra care is needed when bottling your honey for showing.

To get the jars spotless, wash them with washing soda, as detergent can leave a film on the glass which can taint the honey. Let them drip-dry upside down and when dry buff with kitchen towel, especially the jar's shoulders. To reduce air bubbles, pre-heat the jars before pouring in the honey. Slightly overfill each jar so that you can skim off any bubbles that rise to the top and still retain the correct weight. Carefully wipe and clean the rim to stop any honey trickling down the sides. Any bubbles that do appear, commonly round the shoulder of the jar, can be scraped away with a thin, bent piece of wire. Any that settle at the top can be skimmed off with a knife. Save a few extra jars at the time of bottling so that you can choose the best when the date of the show comes around.

There will also be strict rules on the size and content of the labels, so check the entry requirements.

There are different classes for showing honey according to its colour (light, medium or dark); what plant it derives from, and the form it takes, which includes:

- liquid honey (extracted honey with little processing)

- set honey or creamed honey (blended liquid honey where the crystallization process has been controlled by careful stirring)
- comb honey (honey-filled beeswax comb cut off the frame)
- chunk honey (a piece of beeswax comb honey in a jar with liquid honey poured around it).

Honey judges are meticulous and have an array of tools to aid them in their search for the perfect entry. A large torch to shine through the jars, along with a magnifying glass, picks out any impurities that may be lingering in your honey; tasting rods check the flavour and texture; coloured glass identifies the grade of the honey; and a refractometer measures the water content. They also have a weighing

- the information needed on your labels
- the weight of honey you can sell
- the type of container and lids to use.

It is also likely that your region will have specific rules for handling and dealing with foodstuffs for public consumption, so do familiarize yourselves with the basic risks and practices involved by contacting your food standards agency. Your local council or national honey board will be able to advise you on specific criteria.

IS MY HONEY ORGANIC?

Honey can only be called organic if the hive is situated on land that has been certified as organic and since bees can forage quite a distance away from their hive, anywhere within at least a 5km (3 mile) radius of the hive must also be organic or uncultivated. If you don't adhere to those criteria, you can't label your honey as organic, so it is safer to say locally produced. That way people can choose to buy your honey instead of another variety produced further away that has added 'food miles'. Honey is highly nutritious. It contains vitamins and antioxidants and is fat free, cholesterol free and sodium free. Local honey is also thought to be an antidote to hay fever. And urban, polyflower honey, made from the rich biodiversity of flowers and plants found in our towns and cities, is thought to be more flavoursome than honey from the more monoculture countryside.

machine to ensure the weight is exactly as it should be. With their tools and experience, the judges will quickly whittle down the entries to a shortlist of around six jars that will be further examined for smell, taste, weight, clarity, texture and overall quality.

Judges and fellow competitors are a friendly bunch and will be happy to share their tips and techniques, so if your first year isn't a success, take their advice on board. Most importantly, don't give up.

Selling your honey

You may decide that you want to sell your honey at a farmers' market, at your local village fete or even to a local retailer. Whichever outlet you choose you will have to adhere to guidelines for selling food products direct to the public. Each country has its own rules which cover hygiene, food standards, health and safety and labelling. Honey is no exception.

Wherever you live, there are likely to be rules about:

- the water content of your honey, probably limiting it to under 19 per cent, since honey is liable to ferment when the water content is greater than that
- the processes you can employ in production, such as pasteurization

Honey in folklore

In Greek and Roman mythology, ambrosia is the food of the gods and nectar their drink. Both are believed to be honey, which was thought to have an 'ethereal nature and origin'. As Aristotle said, 'Honey falls from the air. Principally at the rising of the stars and when the rainbow rests upon the earth.' The Roman encyclopedist Pliny the Elder saw bees as merely carriers of the food he called the 'sweat of the heavens' and 'the saliva of the stars'. Virgil said that bees 'partake of an Essence Divine and drink Heaven's well-springs'.

Throughout the ancient world honey was used to mark birth, marriages and death, and was a widespread medicine. Babies were fed it to ward off evil spirits and in Greek mythology baby Zeus was thought to have been saved from being killed by his father by eating honey fed to him by sacred bees or bee maidens. The Egyptian pharaohs used it to make offerings to their gods and were buried with it. Their funeral vaults, dating from 1450BC, feature wall paintings of collecting honey, jarring honey and baking honey cakes.

In the Old Testament there are numerous references to the promised land as 'flowing with milk and honey', which symbolized riches and plenty. But honey has also come to be a symbol of the sweetness of love. Cupid's arrows were sometimes said to be dipped in honey, while the Hindu god of love, Kama, carried a bow with a string made of bees. The word 'honeymoon' may come from the Viking custom of a bride and groom eating honeyed cakes and drinking mead for the first month of

their marriage, although it may just mean that the first weeks of marriage are the sweetest.

In the ancient world honey was used for embalming and its antibacterial and antiseptic qualities healed wounds, treated stomach disorders and cured coughs. There were even audacious claims that it could cure baldness and blindness. Today, taking local honey to treat hay fever is gaining credibility.

Ever since Cleopatra bathed in asses' milk and honey, its rejuvenating and beautifying powers have been legendary.

The oldest records of humans taking honey are in prehistoric cave paintings in Spain. One, at Cuevas de la Arana in Valencia, shows a man climbing up a cliff to rob a swarm of wild bees. It is dated around 15,000 years ago, just after the ice age. Honey hunting continued into the ancient, medieval and modern world. The pioneering spirit of the honey hunters in the wild west is celebrated in US honey lore, since it is said that the mouths of the bees' caves were guarded by rattlesnakes, bats, scorpions, ghosts and Cherokee Indians.

Above *An illustration from the 15th-century* Tractatus de Herbis, *by Dioscorides, showing honey collection. The* Tractatus *contains a vast amount of knowledge and information about plants, their virtues and medicinal uses, that is directly descended from the Greek, Roman and Arabic herbal tradition.*

Bees in Winter

The winter months are the quietest for the beekeeper. The bees are huddled together in the hive keeping warm and feeding off their stores of honey. But before the cold weather sets in, they will need your help if they are to survive until spring arrives. As well as ensuring that their home is watertight, well ventilated and guarded against unwelcome intruders such as mice, you need to prepare a large meal to compensate them for the honey you have removed, and practise diligent pest control to prevent your colony being weakened by disease.

without any problems. The decision to take the screen off or not will probably depend on your local climate and conditions, but if in doubt remove it.

Once the above jobs have been done the bees can be left until the spring with little interference from you.

What is happening in the hive

By this time of year, the bees will have reduced the colony size to around 10,000 – a cluster large enough to maintain the heat in the hive but small enough to live off the amount of honey stores. The queen will take a rest from laying, though she will begin again shortly after the winter solstice.

The bees cluster together in a huddle, taking turns to be on the colder edges, and feed on their stores. Since the workers are not out foraging, most of those that were born in late autumn will live through to the spring. There will be some deaths, however, and the bodies will be dragged out of the hive. So it is not surprising to see some corpses on the

> It is not unusual to see bees coming into the hive with pollen on their legs on a warm winter's day. This is a good sign, since fresh pollen is an indication that the queen is laying.

floor outside the hive entrance throughout the winter.

On a cold, sunny day some bees may take a short trip out of the hive to stretch their wings and will take the chance to relieve themselves, since they keep the inside of their home very clean. They won't go far and they may even collect some fresh pollen if there is any nearby. It is not unusual to see bees coming into the hive with pollen on their legs on a warm winter's day. This is a good sign, since fresh pollen is an indication that the queen is laying eggs and the larvae are being fed.

If all goes to plan, the mass of bees will work its way around the hive, eating the stores of honey and begin to increase in population as the winter ends and spring arrives.

Winter feed?

As with most aspects of beekeeping, there is a lot of debate about whether or not you should give your bees a winter feed of candy. Some compare it to junk food because it gives the bees a sugar high, but they can't store the food for later use. Others say an energy rush is better than possible starvation. As a compromise, others suggest you keep the candy as a precaution against bad weather in early spring, when there will be meagre food supplies in the hive. You can buy blocks of candy from bee suppliers. Wrap one in greaseproof paper or newspaper, make a hole on the underside and place it on top of the crown board's feeding hole, leaving some room for ventilation. The bees will come up through the hole to eat it.

Left Even when it's freezing outside, bees keep warm in the hive by clustering together.

Right Bees need an autumn feed as soon as the summer starts to draw to a close. Even if the weather is still warm and sunny, the honey flow will have slowed and the bees in the hive will be preparing for winter.

Stock check

Take advantage of the quiet winter months to stack and store away any frames and supers that are not being used. Store these outside in the cold, but keep them dry and ventilated. The cold will keep away the wax moth.

This is the time to make a note of the equipment you have, to give it a thorough clean, see if any needs replacing and decide whether you will need extra equipment for your second year of beekeeping. Would your next harvest be easier if you invested in an extractor? What about a more light-weight bee suit? You might even be thinking about whether to increase the size of your apiary – two hives are not much more work that one. It is also worth investing in a second hive in case your bees swarm in the spring and you choose to keep the swarm (see page 100).

SCORCHING

The woodwork of beehives is hot and humid – perfect conditions for many unwanted organisms. The most effective way to clean and sterilize anything you are going to reuse and to keep it pest- and disease-free, is to scorch the woodwork with a blowtorch – the type commonly used by plumbers. Do this after the spring cleaning in your second and subsequent years. Thoroughly scorch the insides and outsides of the brood and super boxes, floors, crown boards and queen excluders, paying particular attention to the corners and any small cracks. Don't burn the wood, just give it enough heat to toast it. This operation is especially important if you have more than one hive and you interchange hive bodies and parts.

Second year

When the bees begin foraging regularly for food again, their appearance heralds the beginning of spring as surely as the crocuses and the daffodils. Start your new season of beekeeping by giving the hive a thorough spring clean as soon as the days are warm enough for you to open it up. Your bees have been cooped up in their home for months, so they will appreciate new comb and a new brood box. Bees may swarm in the queen's second year. Regular inspections of the hive will allow you to judge whether or not they are planning to do this and to implement swarm prevention or control. Swarming is the bees' natural way of propagating the species and is nothing to be scared of. It may alarm the neighbours, but a swarm is often easy to collect and is a free and simple way to increase your apiary.

The start of the second year

Hopefully, with your care and support, the bees will have made it through the winter and you will be able to enjoy your second year as a beekeeper.

Early spring

However, don't think you can sit back and relax just yet. The arrival of the first spring flowers can signal a particularly perilous time for bees. The warm weather might bring some welcome flows of nectar and pollen and this will set in motion the bees' natural tendency to increase the size of the colony, making them think the higher temperatures are here to stay. Often, however, especially in temperate regions, a period of cold, rainy weather can quickly follow. This will prevent the bees from foraging and they will have to rely on their dwindling stores to feed the increasing population of brood and young bees. This is the time when the bees are most likely to run out of stored food and starve. For this reason, you must to be prepared for any action you need to take to help your colony during these few weeks. If the weather does change for the worse, feed the bees a strong sugar solution in a contact feeder and refill it if needed. It is better to be safe than sorry. As with the winter feed, you can't overfeed them: if the bees need the food they will take it down into the hive.

First look

On a warm, fine spring day you can have a quick peek to see what's happening in your hive. Check the entrance first, since this can tell you a lot about the wellbeing of the colony. Bees arriving with pollen are a sign of a laying queen and healthy brood. Look for signs of diarrhoea – lines of brownish deposits 2.5cm (1in) long on the landing board and hive walls – this may be due to dampness or to Nosema (see page 122). Put your ear to the wall of the hive and listen for the healthy buzzing. It's a lovely sound.

You can take the roof and crown board off just for a moment. If you see bees on four or five of the frames, this means that all is well.

Don't rush to open the hive in spring. Remember that bees are sensitive to cold weather and the brood needs to be kept at 35°C (95°F). So you will have to be patient and wait until the days are warm enough for you be outside in short sleeves before you open the hive for a spring clean.

Spring clean

The intention of the spring clean is to give the bees' home a freshen up and to replace the old brood comb with new fresh comb, which reduces the risk of disease and swarming. The spring clean takes about a month to complete, but requires only three visits of around half an hour each.

For the first visit, wait for a warm day and plan to be at the hive for the warmest part of it, so that the foragers will be out collecting the fresh spring nectar and pollen. You will need to take a new brood box filled with frames and foundation, a new crown board and a feeder with strong sugar-syrup solution. Light your smoker, put on your protective gear and give the hive a small puff of smoke a few minutes before you open it up.

When you open the hive you should see the bees occupying four or five frames, not necessarily in the middle of the brood box. These will be the frames that have brood on them, with nurse bees attending to the larvae. The other frames are likely to be nearly empty. Those are the ones you want to remove.

Start at the end furthest from the bees and look at the first frame. If it is empty bar a few cells of capped honey, remove it and place it to one side. Carry on until you come across a frame which has pollen on one side – this will probably mean that the other side has brood. If it has, put it back into the brood box in the same position. Now start at the other end of the box and do the same. You will end up with around five frames of pollen, brood and bees. Put these five frames in the centre of the brood box and on either side return a frame that had quite a lot of honey on it. You will now be left with around seven frames centred in the brood box.

Take your new brood box full of new foundation and put this on top of the old box with the frames lined up with each other. Now put the feeder on and fill it with strong sugar-syrup solution. Drip some solution into the hive to let the bees know there is food at the top. It doesn't matter how much solution there is in your feeder, since you will be topping it up over the next week or two. This feed gives the bees the energy to draw out the new foundation. Close the hive.

Once the bees have started to draw out the new, clean foundation the queen will hopefully move up into the new brood box and start laying there, because this process replicates the bees' natural method of keeping their nest clean by migrating upwards and leaving the bottom, used comb, which is then destroyed and removed by vermin and pests.

Over the next few days keep refilling the feeder and after a week, on another fine day, look in the top brood box to see if the queen is laying. If she is, introduce the queen excluder between the two brood boxes. This keeps the queen upstairs. If she isn't in the upper brood box, try smoking the entrance, which may encourage her to move upstairs. If this doesn't work, you will need to find her and drop her into the top box. She will soon be followed by her attendants and the workers. Twenty-one days after the queen has moved into the upper chamber, all the brood in the lower box that she laid just before she

left will have hatched. Time to finish the spring clean.

At this last visit, remove the old brood box and frames, replace the old floor, entrance block and crown board with new (or cleaned) ones and give the hive a new Varroa-mesh floor. You can reuse the old brood box after cleaning and scorching, but dispose of the frames, after melting down the wax, in case they are harbouring any disease.

As a result of your spring clean, the bees can enjoy a hygienic home with a reduced risk of infection, but they also have lots of clean foundation which will keep them busy and thereby reduce their swarming tendencies.

Keep a close eye on the new brood box to ensure the bees don't run out of space. Once a good spring flow is available the colony will expand very rapidly and if they feel their new home, however clean, is too small, they could be off. Keep an eye out for queen cells and be ready to add your first super of the season.

Above *It is a good idea to give your hive a spring clean as soon as the days are warm enough for it to be open. Old foundation can get damp over the winter, and new, fresh comb can reduce the risk of disease.*

Left *To start the spring clean have five frames at the centre of the brood box with pollen, brood and bees, and on either side a frame that has a lot of honey.*

Swarming and increasing an apiary

The honey bees' natural way of propagating is to swarm, creating two colonies from one, which effectively multiplies the species. The old queen and around half the colony leave and settle in a new location. The bees that are left behind raise a new queen, look after the larvae and unhatched pupae and grow to the size of a full colony.

Swarming is healthy for the bees, but not always welcomed by the beekeeper. It not only reduces the amount of honey that a hive produces, since the deserters take a lot with them and there are fewer workers to forage for food, but it is likely to scare the neighbours. A black cloud of 25,000 bees is an alarming sight for most people.

Swarming normally occurs in late spring or early summer. The triggers are:

- overcrowding, which means the queen has too little laying space
- old, unusable comb
- an ageing or diseased queen.

How to collect a swarm

Although swarming can look menacing, as long as everyone stays calm it is a fascinating event. If you are lucky and the swarm lands somewhere accessible, then you can collect it and increase your apiary for free.

> Swarming is heathly for the bees, but not always welcomed by the beekeeper. A black cloud of 25,000 bees is an alarming sight for most people.

When a swarm leaves the hive, usually before midday, the bees fly around in a haphazard fashion for some 20 minutes before massing into a black cloud and finding a temporary place to rest, such as a tree branch, within 50m (50 yards) or so of the hive. Hanging from the branch, they surround their queen in the shape of a rugby ball and wait there for anything from a hour to a day until their scouts have located a viable new home. This sojourn is your opportunity to collect and rehouse a swarm. Filled with three days' worth of honey they are a pretty soporific bunch, but you should still wear your protective gear and have your smoker at hand.

The principle of collecting a swarm is pretty straightforward. You want to get the bees off their temporary landing spot and house them in the new brood box. You do this by either giving the branch they are on a short, sharp shake, or cutting it, so it and the bees fall into the container that your assistant (also wearing protective gear) is holding just below. If the swarm is on a gatepost or fence, you can use a bee brush or your gloved hand to dislodge them; alternatively, if you can get the receptacle above them, you may be able to encourage them into it by giving the cluster a gentle puff of smoke from below.

Once you have them, cover the container with your sheet and take them to their new home. The bees' natural instinct is to follow the scouts and the queen into the new home and you can encourage this by placing them outside the hive and letting them find the entrance. Start by positioning a piece of board (1.2m x 60cm/4ft x 2ft) on the landing board and let it slope down to the ground. Cover it

with your white sheet. Make sure the sheet and board are touching the entrance but not blocking it. If you now turn over the container of bees and dump them on to the sheet, they will crawl up it (they like walking uphill) to the dark entrance. Once the queen is inside the rest of the swarm will follow. It may take 30 minutes for the bees to walk up the sheet and into the hive. This process lets them think that they found their new home naturally and enables them to settle in that bit quicker.

Alternatively, you can just turn the container upside down over the new hive and shake the bees into the new brood box, complete with frames. Put the roof on the hive and leave the skep or box near the entrance. The bees inside the hive will be wafting their

EQUIPMENT

The equipment you are likely to need:

- a pair of secateurs or branch cutters to cut the branch the bees are dangling from
- a receptacle for the swarm, such as a skep, cardboard box or nuc box
- a step ladder if the swarm is high up
- a sheet to cover the receptacle once the bees are inside
- an assistant, as it's difficult to reach up and cut a branchful of bees while holding a box underneath to catch them.
- If you are planning to keep the swarm to increase your apiary, have a brood box, floor and frames in a suitable location so you can transfer the swarm once you've caught it.

Above *If the bees swarm on a gatepost or fence, you can use a bee brush or your gloved hand to dislodge them. Alternatively, if you can get the receptacle above them, you may be able to encourage them into it by giving the cluster a gentle puff of smoke from below.*

Right *Swarms of bees can look menacing – especially to people who are not used to them – but they are simply the bees' natural way of propagating. It is also a way of increasing your apiary, for free!*

scent into the air and the stragglers will soon find their way to the new home.

You now have a second colony of bees. If you decide to keep them, rather than give them to your local organisation or another beekeeper, be aware that they expand very quickly, so be ready to add a super.

If you don't feel comfortable, you don't have assistance or the swarm is inaccessible, contact your local beekeeping association which will have experienced collectors on standby to come and take the swarm away. It will be given to one of the people who have put their name down on a swarm list in the hope of increasing or starting an apiary.

Cast swarms

It is likely that if a colony swarms once the remaining bees will swarm again, and possibly a third or fourth time.

When a colony swarms it leaves behind half the bees, brood at all stages and a new queen in the making. It ensures that there is an emerging queen by rearing a number of them, maybe a dozen. These unborn virgins will emerge nine days after being capped over, which is usually the day the first or primary swarm leaves the colony.

The first virgin queen to emerge after the primary swarm has gone may decide to throw what is called a 'cast' swarm. She takes half the remaining bees, probably around 13,000, and looks for a new home. Those who are left will wait for the next virgin queen to emerge from a capped queen cell and she may cast again by taking half the remaining bees away. This could continue over a 10–12 day period as the queens are born, until the hive has been vacated completely.

Left *A swarm will often collect on a tree branch. They hang from the branch, surrounding the queen and wait there until the scout bees have located a suitable new home.*

Swarm prevention

There are many different ways to stop your bees swarming and none of them is foolproof. But you can use your knowledge of bee behaviour to give them the impression they have swarmed, which satisfies their instinct but keeps them in your apiary.

Because bees often swarm when they are overcrowded, an ongoing swarm-prevention technique is to make sure that the queen has enough brood area to lay her eggs. This means plenty of empty comb, which is why it is a good idea to replace old brood comb with new comb each spring. The colony also needs plenty of room to store nectar, so add supers to the hive before they are needed rather than wait until the storage room has run out.

Further techniques include replacing queens every year or second year. A younger queen's pheromones are strong, so they suppress the queen-rearing instincts of the colony, making it less likely to swarm.

Queen cells

Can you tell when a colony is going to swarm? Since it will only do this when it has reared a number of virgin queens to take over the remaining colony, you need to know what queen cells look like. If they are occupied by a larva this indicates that a swarm is about to happen.

A queen cell is a cone of wax about the size of an acorn, usually hanging downwards from the bottom of the frames, though it can also hang downwards in the face of the frame. It is common to find an empty queen cup

tip
A colony will only swarm when it has reared a number of virgin queens to take over the hive.

which is really a queen cell that is waiting for an egg. If you see a lot of queen cells, your colony has probably started preparations for swarming.

A queen cell comes to a narrow point. If this point has a hinged flap that is open, and by looking into the cell you can see that it is empty, then your colony has raised a virgin and she has emerged. This will probably mean that your bees have already swarmed and you missed it. If this is the case and there are eggs and larvae in your colony, you have a new, young queen in her first year. Let the colony get on with things but make sure they have enough room throughout the year to prevent further swarming. If there is one open queen cell but other closed queen cells you can assume the virgin has just emerged and is in the hive. You will have to destroy the other queen cells to stop the hive throwing a cast swarm. A hive can only support one queen, so if another virgin queen is born, the colony could swarm again.

If, however, the queen cells are not empty but have larvae in them, the colony is getting ready to swarm and will be off as soon as the queen cell is capped. This happens 5 1/2 days after the egg has hatched into a larva, so you have time to prevent the swarm leaving. You do this by splitting the colony, which will give the bees the impression that they have swarmed and curb their instincts for swarming again. One of the many ways to do this is to create a nucleus.

Creating a nucleus

You will need a nuc box or a spare brood box with five brood frames, a floor and cover board.

1. Look through the hive and find a frame with pollen and honey. Place this and the bees into the nucleus box.

> You can use your knowledge of bee behaviour to give them the impression that they have swarmed, which satisfies their instinct but keeps them in your apiary.

2. Find the queen and the frame she is on and place this into the nuc box too. This frame should have a little brood at all stages. If you see any queen cells, remove them.
3. Find another frame with stores of pollen and honey and put this into the nuc box, again removing any queen cells.
4. You need some bees in your nuc box, so shake and brush all the bees off two more frames into the box.
5. Fill the remaining space with empty frames.
6. Close the box and position it on a new stand at least 1m (3ft) away from the original hive.
7. Look through the original hive and destroy any queen cells that are about to be capped, leaving some queen cells that have young larvae in them.
8. Close up the remaining frames in the original hive and fill the spaces with new frames.

   You now have the original hive with queen cells and no queen, and a nuc box with the old queen, some brood, nurse bees and some stores.
9. In one week return to the original hive and destroy all but one of the capped queen cells. The best way to do this is to place the first frame you find with a queen cell to the side, before destroying it – just in case the first one you find is the only one – and then destroy all subsequent cells.

10. Once you are sure there is only one queen cell left, put it back in the hive, close the hive and leave for four weeks. Don't be tempted to open the hive too soon as virgins and newly mated queens are easily disturbed. When you return for the inspection the virgin should hopefully have mated and begun to lay.

The nucleus can be left to build up into a new colony or, if you don't want to expand your apiary, the bees, without the queen, can be united with the original colony when its new queen is settled in.

Has your hive swarmed?

Bees swarm during the day, when you are quite likely to be out. When you come to inspect your colony you may notice that they are a bit more agitated than usual and buzzing loudly. This may be an indication that they do not have a queen who is laying eggs. Look through the colony to see if there are any empty queen cells. If there are, then your bees have probably swarmed. To make sure, check for eggs and young larvae. If you can't find these, then your laying queen has gone. There will probably be a virgin queen running around the frames. She will be difficult to spot since she is smaller than a mated queen (her spermatheca is not full and expanded). Try to find any other queen cells and destroy them. A cast swarm will only happen if there are queen cells in the colony.

Close the hive and leave it for four weeks to give the virgin time to mate and start laying. When you inspect the hive after a month you should find new brood, which indicates that the colony is back on track.

Bee products

Not only will your bees give you honey, they will also produce beeswax with which you can make the most heavenly scented furniture polish and candles, and natural cosmetics that are thought to have special health-giving powers. After extracting the beeswax, you can have hours of fun with all the family learning to create different types of candles in all shapes and sizes, or making a variety of creams and ointments. It can be a bit fiddly at first, so don't be put off if you don't get it right straight away. Your patience will soon be rewarded and you will enjoy yet another contribution from your bees. Once you have mastered these crafts your artistic work will make ideal gifts for all occasions.

Beeswax

Your wax will come from the cappings that you have saved during the honey harvesting as well as scrapings of comb that have come off the frames and hive during the course of the year. In your second year you will have more wax as you replace some of the older frames with new foundation. This wax will have to be melted and filtered to give you clean beeswax for your candles.

Extracting beeswax

As your apiary grows you may wish to invest in a solar extractor, which is a clean and labour-free way of getting wax off the frames. It consists of a shallow box, into which you place the frames, with a glass lid which traps the sun's heat, causing the wax on the frames to melt. The wax is collected in a tray after passing through a filter which collects the debris.

If you don't have a solar extractor you'll have to melt the beeswax by hand. Utensils that come into contact with wax will be covered with the stuff, so have dedicated ones just for wax handling.

You will need a double-boiler or bain marie. This is a pot within a pot that allows you to heat the wax indirectly by boiling water in the outer pot while using the inner one for the wax. You can improvise by placing a small pot into a larger pot of boiling water, or a glass or stainless steel bowl over a large container of water.

You will also need a metal strainer and collecting pan for coarse straining, and nylon tights or muslin to fine-strain the wax.

Warning: Never melt wax directly on a flame or in a microwave. It gets hot too quickly, and can easily burn and ignite. Always use the double-boiler method. Make sure all workbenches are covered with old newspaper to catch the inevitable drips.

1. Heat the water in the larger, outer pot and place the collected wax into the inner pot.
2. Add some water to the wax so that it has something to float on and debris contained in the wax will sink to the bottom.
3. When all the wax has melted, pour it through the metal strainer into the collecting pan. This gets rid of larger bits of waste such as old silk cocoons, bee parts, propolis and even the odd bit of foundation wire.
4. Give the warm inner pot a quick clean with some kitchen towel and return the strained wax to it. Melt the wax for a second time and strain it through the tights or muslin to get rid of finer particles such as pollen.
5. Pour it back into the inner pot with enough rainwater to keep the wax off the bottom. Heat it for a third time until the wax has melted again. Now let it cool for a day or two. It shrinks slightly on cooling and breaks away from the edges of the pot, leaving a cake of yellow wax floating on top of the water.
6. There may be some bits that need scraping off the underside of the wax to leave a perfect block of clean wax.

Put aside some beeswax in a small plastic container for use around the home. It makes a marvellous polish that not only seals the wood, but creates a high sheen and gives off a lovely aroma.

Beeswax wood finish (for bare wood)

You will need: 55g (2oz) beeswax, 300ml (½pt) turpentine, bain marie

1. Melt the wax in the bain marie.
2. Remove from the heat source and pour in the turpentine.
3. Return to the heat and give a gentle stir.
4. Cool, and spoon into sealable containers.
5. Warm slightly before applying to bare wood.

> ## tip
> Beeswax obtains its colour and fragrance from the propolis and pollen brought into and stored in the hive.

BEESWAX FACT BOX

Beeswax is a by-product of the honey bees' honey consumption. It takes about 4kg (8 1/2lb) of honey to produce 450g (1lb) of beeswax.

Worker bees aged between 12 and 18 days secrete beeswax in the form of a scale about the size of a pinhead from wax-secreting glands under their abdomen. It's estimated that it takes about 500,000 scales to make 450g (1lb) of beeswax.

The secreted wax is tasteless, odourless and almost colourless. It obtains its yellow colour and its fragrance from the propolis and pollen brought into and stored in the hive. Beeswax has a melting point between 63 and 65°C (146 and 149°F).

Apply with a soft cloth, allow to dry, then buff with a wool cloth. This product can also be used on leather.

Beeswax wood conditioner (for finished wood)

You will need: 55g (2oz) beeswax, 600ml (1pt) turpentine, 2 cups water, 25g (1oz) soap flakes

1. Shred the wax into the turpentine, cover and leave in a warm spot such as a sunny windowsill for a few hours, shaking from time to time until the wax has dissolved.
2. Bring the water to the boil, add the soap and stir until dissolved.
3. Let this cool and add it to the wax/turpentine mixture, stirring briskly until an emulsion forms.
4. Pour into sealable glass bottles and shake well before use.

Dubbin (traditional waterproofing for leather boots and shoes)

Stir 110g (4oz) beeswax and 55g (2oz) of palm oil together in the bain marie until the wax has melted. Let it cool slightly before pouring into a jar. Warm slightly before applying to leather.

Above *Beeswax is a highly-prized polish for bare and finished wood, which gives a high sheen and a lovely smell.*

Other uses for beeswax: filling in cracks and holes in furniture; rubbing on thread for easier needle threading and sewing; lubricating the runners on sticky wooden drawers, saw blades, woodworking tools and zippers on clothes. Fishing lines float better with a coating of wax.

Beeswax candles

There is something special about a soft glowing light from a candle enhanced by the delicate perfume of beeswax. Making your own candles is a skill that requires time and patience to master, but if your first efforts aren't perfect, you can always melt them down and try again.

Moulds

Moulds can be made from glass, metal, plastic and silicone and are available from craft shops and online stores. In your first year, when you may have little wax, try making tea lights using old tea light containers. You can also use old yoghurt pots.

The wick

The secret of a good candle is in the wick. You need the correct size in relation to the diameter of the candle – too small and the candle won't burn, too large and it will smoke. Beeswax candles need a wick about twice the thickness of that for paraffin candles. If you are buying wick from a beekeeping supplier they will sell it for beeswax, but a general candle-making shop will probably only sell wick for paraffin candles. A bit of trial and error will get you the right wick-to-candle ratio.

The wick then needs to be primed so that it is easier to light and thread through the hole in the candle. Simply cut it to length and immerse it in some hot wax. Once the wick has had time to absorb the wax, remove it, stretch it straight and place it flat to let it dry. This will only take a few minutes.

You will need: a yoghurt pot mould, wick, molten beeswax, scissors, a little dishwashing liquid or glycerine

1. Make a small hole in the bottom of the pot and thread the wick through the hole. To straighten it, tie a knot in the hole end and pull it tight. Wrap the other end around a toothpick or pencil and lay this across the opening, making sure the wick is centred.
2. Take your block of clean wax and melt it in the clean bain marie.
3. Line the inside with the dishwashing liquid or glycerine, which will allow the candle to break free from the mould once dry. Silicone rubber moulds do not need this.
4. Block the hole in the base with a little mould seal (Blu-Tac) to stop the molten wax escaping from the hole.

tip
The secret of a good candle is in the wick. Beeswax candles need a wick about twice the thickness of that for paraffin candles.

5. Position the mould on a tray, slowly pour the molten wax into it and let it cool for 24 hours.

6. Cut the knot that was holding the wick in place and you should find that the wax has shrunk and falls easily out of the mould.

7. Flatten the bottom of the candle by heating the double boiler and rubbing the candle on the bottom of the inner pot until it has a flat and smooth end. Your candle is now ready for lighting.

Foundation candles

The foundation in the frames can also be used for making home-made candles. These are simple to make, burn well and are a great alternative to moulded candles when you run out of supplies of wax cappings.

The foundation sheets must be the unwired variety. They are sold by beekeeping suppliers and come in a huge variety of shades and bright colours for candlemaking.

1. Foundation works best when it is warm – it gets brittle when it is cold – so if necessary place it in a warm spot to soften it.

2. Measure the wick against the long edge of the foundation and cut it about 15mm (1/2in) longer, leaving enough sticking out the top for lighting.

3. Lay a sheet of foundation on a flat surface and place the wick along the edge to be rolled. Fold the foundation tightly over the wick and press to create a straight edge. Now continue to roll the wick, like a Swiss

Roll, keeping the cylinder straight and tight.

4. When the whole sheet is rolled, stick the loose edge down by gently rubbing your finger along the edge. You should now have a pure beeswax candle ready for lighting.

Variation: An attractive tapered candle can be made by using sheets of foundation that have been cut diagonally. When folding, keep the longer, straight edge as the base and the candle will roll into a conical shape. To add to the fun, you can decorate your candles by

adding shaped bits of coloured wax. Cut out stars from foundation sheets, for example, and press them on to the candle. Use a little molten wax to glue them on if necessary.

Below Foundation candles are easy to make using different coloured sheets, and make ideal gifts. Because no molten wax is involved, they are also perfect for children to make.

Cosmetics

Beeswax has been an essential ingredient in beauty products for centuries, sought after for its natural softening and protecting qualities for skin. Balms, creams and ointment can all be made quite simply in the kitchen by mixing little more than essential oils and water into your own beeswax. Craft shops stock a range of attractive containers that you can use for homemade cosmetics.

Beeswax lip balm

This is an amazingly easy recipe.

You will need: 30ml (2 tbsp) grated beeswax, 20ml (4 tsp) almond oil, 10ml (2 tsp) honey, a few drops of an essential oil (optional), double boiler, small containers

1. Put the ingredients into your double boiler and simmer over water until they have melted.
2. Remove from the heat and stir together.
3. Add a few drops of your favourite essential oil if you like.
4. Pour the liquid into small containers.
5. When the mixture is cool put the lids on.

Milk and honey bath

This is a modern take on Cleopatra's bath of asses' milk and honey.

You will need: 90g (3oz) powdered milk, 60ml (4 tbsp) honey, bowl

1. Mix the milk and honey into a paste.
2. Dissolve in a warm bath as you run it.

Body lotion

A light lotion with a fragrant smell that makes an ideal present.

You will need: 300ml (½ pint) aloe vera gel, 5ml (1 tsp) lanolin, 5ml (1 tsp) pure vitamin E oil, 100ml (3 ½fl oz) coconut oil, 15g (½ oz) beeswax, 225ml (8fl oz) almond oil, up to 7.5ml (1 ½ tsp) essential oil of your choice (or more to prolong the scent), blender, double boiler, glass jars

1. Put the aloe vera gel, lanolin and vitamin E oil in a blender.
2. Put your coconut oil and beeswax in the double boiler and heat until fully melted.
3. Into this melted wax stir the almond oil, reheating if necessary until mixture has blended together.
4. Run blender at low to medium speed, then pour in melted oils in a thin stream. As the oils are blended in, the cream will turn white and the blender's motor will begin to grind.
5. When you have a mayonnaise-like consistency, stop the blender and add the essential oil.
6. Transfer the cream to glass jars while still warm because it thickens quickly.

Coconut butter body moisturizer

A very rich moisturizer good for dry or weather-beaten skin and sunburn.

You will need: 30ml (2 tbsp) beeswax, 10ml (2 tsp) distilled water, 110g (4oz) cocoa butter, 60ml (4 tbsp) sweet almond oil, 30ml (2 tbsp) coconut oil, double boiler, glass jar

1. Melt the beeswax with the water over a low heat.
2. Remove from the heat, spoon in cocoa butter and blend.
3. Gradually blend in the oils.
4. Pour the mixture into a glass jar. The lotion will thicken as it cools.

Simple hand cream

A quick and easy-to-make hand cream that will soothe and protect your hands.

You will need: 55g (2oz) beeswax, 300ml (½ pint) sweet almond oil, 300ml (½ pint) water, 10 drops essential oil (if desired, for fragrance), double boiler, saucepan, blender

1. Heat the beeswax and sweet almond oil in the double boiler until the wax melts.
2. In a saucepan, heat water until warm. Both mixtures should be warm, but not so hot as to be uncomfortable to the touch.
3. Place the warm water in the blender and

turn it on to high speed.

4. Slowly pour in the beeswax-oil mixture in a thin stream. When most of the oil has been added, the mixture should begin to thicken.
5. Add the essential oil if desired.
6. Turn off the blender when you have a thick cream.
7. Spoon into ointment jars or small metal tins.

Rich hand cream

A thick, enriched hand cream that will relieve and protect dry and chapped hands.

You will need: 225ml (8fl oz) sweet almond oil, 100ml (3 ½fl oz) coconut oil or cocoa butter, 5ml (1 tsp) lanolin, 15g (½oz) grated beeswax, 200ml (7fl oz) rosewater, 150ml (¼ pint) aloe vera gel, 1–2 drops rose blend essential oil, 5 drops vitamin E oil, double boiler, jar

1. In the double boiler melt sweet almond oil, coconut oil or cocoa butter, lanolin and grated beeswax together over low heat and cool to room temperature.
2. When cool, add the other ingredients.
3. Whip the mixture to a smooth consistency.
4. Store in a covered jar.

Above *Beeswax has been used for centuries in a wide range of beauty products, most of which are easy to make at home.*

Recipes

Honey is nature's natural sweetener. It is more delicate, subtle and varied in taste than sugar. It has been used in cooking for centuries – the first known cookbook from ancient Rome used honey in all types of recipes, savoury as well as cakes and desserts. Honey goes particularly well with dairy produce such as cheese and yoghurt. The following recipes reflect its versatility in the kitchen, where it can as easily be used as a marinade for meat, a glaze for vegetables and a sauce for fruit. At the end of the summer you may have more honey than you know what to do with, so why not make some traditional honey cakes that can be stored throughout the year.

Roasted figs with Gorgonzola and honey

This is an exotic-looking and -tasting dish that makes quite an impression served as the first course at any dinner party. However, it is deceptively simple and quick to make. If you don't have Gorgonzola, any creamy blue cheese will work.

- 12 ripe figs
- 160g (6oz) Gorgonzola
- salt and pepper
- 30ml (2 tbsp) honey

Preparation time: 10 minutes

Cooking time: 8 minutes

Serves: 6

1. Pre-heat the grill to its highest setting.
2. Halve the figs and place them cut side up on an oiled baking tray.
3. Season with salt and black pepper and grill for 5–6 minutes, until they are soft and starting to bubble.
4. Chop the cheese into 1cm (1/2in) cubes.
5. Remove the baking tray from the grill and put the cheese on the top of each of the figs. Gently press the cheese down to squash it in a bit.
6. Grill for a further 2 minutes, until the cheese is bubbling and browning slightly.
7. Serve immediately with the honey poured over, and a few drops of balsamic vinegar for presentation.

Honey-basted lamb

Cooked with cider and herbs, this is the perfect main course for Sunday lunch. Serve with garden potatoes and fresh mint sauce.

- just under 2kg (4lb) spring lamb (leg or shoulder)
- 2 tbsp rosemary
- 1 tsp ginger
- 225g (8oz) honey
- salt and pepper
- 300ml (½pt) cider

Preparation time: 10 minutes

Cooking time: about 2 hours (allow 25 minutes per pound and 20 minutes extra)

Serves: 4–6

1. Preheat the oven to 225°C/430°F/gas mark 7.
2. Line an ovenproof dish with foil.
3. Prepare the lamb by rubbing with salt, pepper and ginger.
4. Place the lamb in the dish and sprinkle half the rosemary over it.
5. Pour honey over the top of the meat and cider around it, then put in the over.
6. After 30 minutes, reduce cooking temperature to 200°C/400°F/gas mark 6.
7. Baste during cooking, adding more cider if necessary.

Spicy chicken wing marinade

The aromas wafting across the garden as this marinated chicken sizzles on the barbecue on a warm summer's day are truly mouthwatering. It also has a sensational, rich taste as all the flavours have soaked into the meat.

1. The day before you want to serve this dish, whisk together the marinade ingredients.
2. Place the chicken wings in a large container, pour the marinade over them, cover and refrigerate overnight.
3. Place the marinated chicken on a hot barbecue and cook for 5–7 minutes, turning and basting frequently until cooked through but tender.
4. As an alternative to the barbecue, pre-heat the oven to 190°C/385°F/ gas mark 5 and bake the chicken wings in the marinade for 20 minutes or until cooked, turning and basting frequently.

- 60ml (4 tbsp) sesame oil
- 60ml (4 tbsp) white wine vinegar
- 45ml (3 tbsp) crunchy peanut butter
- 45ml (3 tbsp) clear honey
- 30ml (2 tbsp) chilli sauce
- 900g (2lb) chicken wings

Preparation time: 12 hours
Cooking time: 20 minutes
Serves: 4–6

Mustard and honey salad dressing

This popular sweet salad dressing is easy and quick to make. The hot mustard combines well with the sweetness of the honey and the sharpness of the lemon to create a rich-tasting dressing that goes with most salads.

- 5ml (1 tsp) clear honey
- 5ml (1 tsp) Dijon mustard
- 30ml (2 tbsp) lemon juice
- pinch of salt
- 60ml (4 tbsp) sunflower oil or a mixture of sunflower and olive oil

1. Mix the first four ingredients together.
2. Add oil and whisk together.
3. Pour into a jug.

Amber shallots

These deliciously sweet shallots go well as a side dish with meat dishes. They are easy to prepare but make sure you leave enough time to cook them so that they are really tender.

- 8 shallots
- 1 tsp salt
- ½ tsp paprika
- 30g (1oz) butter or margarine, melted
- 150ml (¼ pint)

tomato juice
- 45ml (3 tbsp) honey

Preparation time: 15 minutes

Cooking time: 40 minutes– 1 hour

Serves 8

1. Peel the shallots and half any large ones.
2. Place in a large, shallow, greased casserole.
3. Combine salt, paprika, melted butter, tomato juice and honey.
4. Pour sauce over onions.
5. Cover dish and bake at 150°C/300°F/gas mark 3 for 1 hour or until the onions are tender.

Honey-glazed vegetables

Ideal with a roast dinner. The delicious honey juices combine beautifully with the meat, but also go perfectly with a nut roast. You can really use any vegetable you like, rather than just those listed here.

- 2 parsnips
- 2 carrots
- 2 red onions
- 30g (1oz) butter
- 30ml (2 tbsp) honey
- 30ml (2 tbsp) lemon juice

- 250ml (9fl oz) vegetable stock
- salt and pepper

Preparation time: 5 minutes

Cooking time: 30 minutes

Serves: 4

1. Pre-heat the oven to 200°C/400°F/gas mark 6.
2. Peel and prepare the vegetables. Cut in half lengthways, then slice into 1cm (½in) pieces.
3. In a large flat pan, gently soften the onion and then remove.
4. Combine carrots, butter, stock and a pinch of salt. Bring to the boil, lower the heat to medium, and cover and simmer for 4 minutes.
5. Add the parsnips and cook for about 5 minutes or until the vegetables are tender. Now return the onion to the pan, and transfer the vegetables to a shallow baking dish with a draining spoon, preserving the liquid in the pan. Add salt and pepper.
6. Make sure you have 250ml (9fl oz) of stock remiaining (adding more if necessary). Stir in the honey and lemon juice. Bring to the boil and simmer for 1 minute or until the liquid has thickened slightly. Pour over the vegetables.
6. Cook uncovered for 15–20 minutes or until lightly glazed.

6. Divide the mixture equally between the prepared tins.
7. Bake in the oven for about 1 hour or until a toothpick inserted into the middle comes out clean.
8. Cool for 10 minutes before removing the cakes from the tin.
9. Leave the cake to cool further on a wire rack.
10. Allow to mature for a day or two before eating.

Honeyed yoghurt breakfast

There is no better breakfast than honey dripped over rich yoghurt. It is nourishing and delicious as the sweetness of the honey offsets the sharp tang of the yoghurt. Add a slice of chopped melon or a chopped-up banana for your morning portion of fruit.

• 150ml (½ pint) thick Greek yoghurt	• 30ml (2 tbsp) clear honey
• 15g (½oz) toasted flaked almonds	
• 25g (1oz) raisins	
• 1 banana or slice of melon chopped	**Preparation time:** 2 minutes **Serves** 4

1. Mix together yoghurt, almonds and raisins in a bowl.
2. Chop up a banana or pieces of melon and add to the mixture.
3. Transfer to a glass serving bowl and drizzle with honey.
4. Serve immediately.

Traditional honey cake

Honey cake is the oldest cake. It was made in biblical times and there are pictures of Egyptians baking it in pharaohs' tombs. It became spice cake and gingerbread in the Middle Ages and is still enjoyed today. It was often made at the end of the summer when the honey and grain were harvested and is still traditionally served at this time as part of the Jewish New Year celebrations to wish for a sweet year ahead. It is a cake to accompany a cup of tea, rather than a dessert after a meal.

• 400g (14oz) plain flour	• 450ml (1 pint) honey
• 175g (6oz) firmly packed dark brown sugar	• 60ml (4 tbsp) vegetable oil or melted butter
• 1 tbsp baking powder	• 200g (7oz) coarsely chopped nuts (optional)
• 1 tsp baking soda	
• 2 tsp cinnamon	
• ½ tsp each ground allspice, nutmeg, ginger and salt	**Preparation time:** 30 minutes **Cooking time:** 1 hour
• 4 eggs	**Serves:** 10–12

1. Preheat the oven to 175°C/350°F/gas mark 4.
2. Grease and flour two 22 x 12cm (9 x 5in) loaf tins.
3. Stir all dry ingredients together until well combined.
4. Make a well in the centre of the mixture and add eggs, honey and oil.
5. Stir mixture until smooth.

Mead

Dating back more than 3,000 years, mead has been enjoyed by the great civilizations of the world. It is also called honey wine because it's alcohol content is similar. It tastes like sherry and, like wine, it matures well. Your guests will be very impressed if you serve it at a party – although it can be an acquired taste!

You will need two large fermenting vessels with airlocks, and a siphon or funnel, all of which you can buy from a wine-making supplier.

- 1.4kg (3lb) honey
- 4 litres (7 pints) water – preferably filtered rainwater
- 1 package wine yeast

Preparation time: 1 month
Cooking time: 1 year
Makes: 4.5 litres (1 gallon)

1. Heat about half the water to boiling point in a large, clean, stainless steel pot.
2. Take the water off the boil, and add about two-thirds of the honey, reserving 450g (1lb) for later sweetening. Stir until the honey dissolves. Add the remaining water. Heat again and remove any scum that may appear. Cool this must (the mixture of honey and water) to around 27°C (80°F).
3. Add the yeast and agitate vigorously to mix it in and to give the must lots of oxygen.
4. Pour this mixture into your fermenting vessel.
5. Install the airlock, place the mead in an area that will maintain a more or less constant temperature of 21–27°C (70–80°F) and wait, making sure the airlock has liquid in it at all times. You should notice the airlock start to bubble in a day or so and continue to do so for about a month.

6. After about a month when the bubbling has stopped, siphon the must into a clean fermenting vessel and airlock again. At this stage taste the liquid and add sterilized honey water if you would like to sweeten the mead. (To make this, stir 1 part honey into 2 parts boiled water and let it cool before adding it to the mead via a siphon or funnel.)
7. Watch your airlock, and when the bubbling has slowed to less than 1–2 bubbles per minute, you can either store the mead for bulk aging or siphon it into sterilized bottles. Beer bottles with caps are best since they can withstand any pressure that may build up if the mead continues to ferment. Wine bottles with corks can be used if the fermenting has truly finished.
8. Leave for a year before tasting.

You can vary the flavour of mead by adding herbs and spices to the recipe. Cloves, cinnamon, marjoram and oranges are just a few of the flavours that can be added.

Health and care

Good hygiene is essential for the health of your bees and is easily achieved by keeping your apiary clean and tidy and not interchanging equipment and components between hives. Sometimes, however, ailments can occur in even the cleanest apiary, since bees are susceptible to a number of parasites and diseases. By taking sensible precautions against the most common pests and following an integrated pest-management system you should be able to maintain strong, vigorous colonies that can use their own defences to ward off foes.

Prevention and cure

It is unfortunately a fact of nature that honey bees are susceptible to a number of parasites and diseases, both in the wild and in the hive. As a beekeeper you will come across the common pests, such as wax moth and Varroa, but hopefully the more serious foulbrood disease will be something that you only read about.

There are a number of different approaches to helping bees deal with their maladies, but the most important is taking preventative action to keep your colonies strong. Strong colonies use their own natural defences to combat their adversaries and diminish the need for you to act as a doctor.

Below *Keeping your apiary clean and tidy and ensuring your colonies are strong will help the bees fight disease.*

Preventative measures

- Keep your apiary clean and tidy.
- Only buy bees that are from disease-free, reputable apiaries.
- Dispose of old comb and propolis rather than leaving it around the hive.
- Don't buy second-hand frames since they can't be effectively sterilized.
- Sterilize, by scorching, second-hand hives – or avoid buying them.
- Don't interchange hive parts and equipment between one apiary and another.
- Combine weak colonies. One strong colony is better than two weak ones.
- If you have a number of colonies prevent bees robbing (stealing an other hive's honey) and drifting (moving into a neighbouring hive) by keeping your entrances small enough to defend and facing different ways so the bees remember which hive is theirs.
- Keep up-to-date with any new methods of pest control and follow an integrated pest-management (IPM) routine throughout the year.
- Keep good records so that you can monitor changes in the hive and know what medication you have administered.
- If your bees die, close the hive to prevent other bees getting access to the honey and picking up the disease.

Warning signs

Unlike with pets, it is hard to tell if your bees are off colour by their behaviour. With experience you will be able to spot some changes in their mood, but the most reliable

INTEGRATED PEST MANAGEMENT

The term 'integrated pest management' has become more common in beekeeping since the Varroa mite became so widespread. Instead of depending on manmade chemicals to fight parasites and disease, IPM takes a more holistic approach, using more natural methods to enable the bees' own defence mechanisms to create a strong colony. One example is dousing bees with fine icing sugar throughout the season (see page 123), a seemingly strange practice that encourages them to groom each other and in the process pick off the little mites that are clinging to their body. Your local beekeeping association will have more details of IPM techniques and will run courses on IPM systems that are recommended by the national body that oversees animal husbandry where you live.

way to assess the health of a colony is to check what is going on in the brood box. A healthy-looking brood box has frames with a consistent pattern of brood laying, pearly white larvae where the segments are clearly visible, uniform brood cappings of a light brown colour, lively bees with perfect wings that seem to be flying well, stores of fresh-looking pollen and white-capped honey cells. There should be no unpleasant smells and the hive should be free from damp and mould. If, during your regular hive inspections, you see that the colony isn't performing as above, seek assistance from an experienced beekeeper, who will be able to identify the problem and advise on a course of action.

There are a number of ailments that effect the bee brood, as well as adult diseases and parasites, and pests. If you suspect your hive has any of these, contact your local beekeepers' association or your local bee inspector for help. In the UK two diseases – American foulbrood and European foulbrood – and one pest – small hive beetle – are notifiable, which means that if you suspect an outbreak of any of these you must, by law, inform the Central Science Laboratory (CSL) at the National Bee Unit (see page 128) and submit a suspect disease sample for analysis.

> *IPM takes a more holistic approach, using more natural methods to enable the bees' own defence mechanisms to create a strong colony.*

Brood diseases
American foulbrood (AFB) and European foulbrood (EFB)

The geographical names of these two rare but serious diseases caused by bacteria derive from where they were discovered, but somewhat confusingly they can occur in any country throughout the world. If your brood is not looking pearly white with the older larvae curving around into a C shape within the cell, this is the first sign that something is wrong.

If the capped cells are sunken and dark in colour, suspect AFB. The test you can administer is called the 'ropiness' test. Insert a matchstick into the sunken cell and twist it. If when you pull it out there is a mucus-like thread attached, then your colony has AFB. There will also be an unpleasant smell. Call your regional bee inspector or local association immediately for help. If your analysis is correct, your hive and the bees will have to be destroyed by burning.

With EFB, you will see larvae that look as if they have melted and are turning a yellowish-brown colour. Again, call for help immediately. EFB can be treated if it is a small infection.

Small hive beetle

The small hive beetle (SHB) originates in Africa, where the native bees *Apis mellifera capensis* and *Apis mellifera scutellata* can deal effectively with their natural predator. The European honey bee, however, can't cope with this tropical pest, which has wreaked havoc in apiaries across the south-eastern states of the USA. The dark brown, nearly black, oval beetle is about 7mm (1/4in) long and 4mm (1/6in) wide and lives and breeds in the colony of honey bees. Females lay masses of eggs. The larvae feed on the bees' brood, pollen and honey. The mass of larvae often congregates together. They don't produce silk threads that looks like webbing (it's the larvae of the common wax moth – see below – that do that). When you open an infected colony you will see the beetle scarper to the dark corners of the hive and its white, three-legged larvae will be all over the brood, eating it. If you suspect your colony has been infected by SHB, contact your regional bee inspector or your local association immediately. These beetles probably enter the country by container, so if you are located near points of entry such as major airports or ports be extra vigilant.

Chalkbrood

A much more common but less serious brood disease is chalkbrood, which is caused by a fungus which germinates in a stressed colony. This can occur when there is a sharp drop in temperature after a warm spell in spring. The fungus grows and is ingested by the larvae, which die shrunken and hard. The chalk-white remains can easily be seen on the hive's landing board as the house bees try to clean out the dead larvae. A strong colony can live with a small amount of chalkbrood, since the cluster can keep most of the brood warm and only the fringes get cold. However, if the colony weakens and can't keep warm, the brood may succumb to the disease. Cleaning out the brood frames and replacing the existing queen with a healthier one will help if the problem continues.

Adult diseases and parasites

Nosema

Nosema is caused by a parasite (*Nosema apis*) which lives in the gut of the adult bee. It gives the bee stomach ache and is passed to the rest of the colony via the faeces. Nosema can only be diagnosed by examination under a microscope, but symptoms include diarrhoea – dark-coloured excrement – on the outside of the hive and possibly on the frames. It weakens the bees and reduces their life expectancy, so the colony becomes depleted. Nosema can be treated by an antibiotic mixed in with the autumn feed. Going into winter with a healthy, vigorous colony with good honey stores will help to keep Nosema down. Replacing old brood frames in the spring should also reduce the incidence of the parasite as the infected comb is destroyed.

Strong colonies use their own natural defences to combat their adversaries and diminish the need for you to act as a doctor.

Varroa

Varroa (*Varroa jacobsoni*) is a mite that lives on both the brood and the adult bees. If it is left untreated the colony will die. This seems to have become a worldwide problem and Varroa is becoming resistant to some of the treatments available to the beekeeper. The reddish brown mite, about 1.5mm (1/20in) long, is easily seen on the white larvae. Careful integrated pest management which uses a mixture of chemical and non-chemical treatments, such as having a Varroa screen floor and removing drone brood, which the mites are particularly partial to, is recommended. Ask your local association, regional bee inspector or an experienced beekeeper for their advice on this common problem.

Tracheal mites (Acarine)

The mite *Acarapis woodi* lives in the breathing tubes of the adult bee which causes premature death, usually in winter, and leads to 'spring dwindling' in the colony. The expanding brood is left unattended by the drop in the adult population and the colony is severely weakened. Only diagnosed by microscopic examination, the symptoms can be deformed wings and weak bees with distended bodies. There is no effective treatment as yet. Keeping strong, vigorous colonies and good husbandry are recommended. There is evidence that the chemicals used for Varroa are also having an effect on this mite. Keep up-to-date with current treatments by talking to your local bee association.

Pests

Wax moth

Wax moth (*Galleria mellonella*) and their larvae are a nuisance but can be kept under control by the combined efforts of bees and beekeeper. The moth lays eggs in the hive and as they grow the caterpillars feed on the wax, pollen, honey and larvae. In a strong colony the bees' own police force will deal with this threat by chasing the moth and the larvae out of the hive, and you can easily see and remove them too. But you won't be able to see the eggs, so in a weaker colony the moth may take over and eventually the colony will vacate the hive, looking for a new home.

Moths' eggs can stay in the hive's boxes and frames even when they are not being used, but they are susceptible to the cold (less than 2°C/36°F), so you can kill them by leaving hive components that aren't being used outside over the winter. If you live somewhere where the temperature doesn't fall this low, put the components in a freezer instead. Don't leave old comb and wax around the apiary, since the wax moth will surely find it and take hold.

Colony collapse disorder

Colony collapse disorder (CCD) has been dubbed a mystery plague. It seems to have been most prevalent in the USA among commercial apiarists. There are as yet no definitive answers to this problem, which has led to bees vanishing from hives. Research is ongoing, chiefly in the USA. The best approach is to keep up to date with developments through your beekeepers' association or beekeeping journals. If you suspect CCD, contact your association or regional bee inspector immediately.

Predators

Bears

In some parts of the world bears can be a problem for your apiary. They like to eat bee brood as well as the honey and will cause much damage to the hives in the process. If you live in an area with bears, keep your hives away from its paths, ravines and forests.

Skunks

Strong bee colonies can overwhelm this insectivore, but weaker colonies might lose out. Protection by fencing and stands can prevent the skunk from scratching at the entrance of the hive. When the bees come out to defend their home they fall prey to the hungry animal.

Badgers

Badgers are keen on honey too. They tend to push hives over by getting underneath them and doing a press-up with their powerful front legs which then topples the hive. Many common hive stands are the perfect height for badgers, so if you are in a badger area you'll need to put up a barrier to stop them getting to the hive.

Below *Dousing your bees with icing sugar is part of varroa control.*

tip
If your bees die, close the hive to prevent other bees getting access to the honey and picking up the disease.

Glossary

Abdomen
The posterior or third region of the body of the bee that encloses the honey stomach, stomach, intestines, sting and reproductive organs.

Acarapis woodi
A mite, also called the tracheal mite, which infests the bees' breathing or tracheal system.

American foulbrood (AFB)
A **brood disease** of honey bees caused by the spore-forming bacterium, *Bacillus larvae*.

Anaphylactic shock
Constriction of the muscles surrounding the bronchial tubes of a human, caused by hypersensitivity to venom (including bee stings). Requires immediate medical attention.

Apiary
The location and total number of hives and other beekeeping equipment at one site.

Apiculture
The science and art of raising honey bees.

Apis mellifera
A native European bee that is kept for its honey and wax in most parts of the world. Commonly known as the European or western honey bee.

Bacillus larvae
The bacterium that causes **American foulbrood.**

Bee brush
A soft brush or whisk (or handful of grass) used to remove bees from frames.

Bee escape
A device constructed to permit bees to pass one way within a hive, but prevent their return; used to clear bees from **supers.**

Beehive
A box or receptacle with movable frames, used for housing a colony of bees.

Bee space
A space big enough to permit free passage for a bee, too small to encourage comb building and too large to induce the bees to produce propolis; for *Apis mellifera* it measures 9.5mm (just over 1/3in).

Bee suit
Protective coveralls to protect from stings and to keep clothes clean.

Bee veil
A cloth or wire netting for protecting the beekeeper's head and neck from stings.

Beeswax
A waxy substance secreted by bees through special glands.

Black scale
The 'dried down' appearance of a larva or pupa which has died of a **brood disease.**

Brace comb
A bit of comb built between two combs to fasten them together, between a comb and adjacent wood, or between two wooden parts such as top bars.

Brood
Immature stages of bees not yet emerged from their cells; the stages are **egg, larva** and **pupa.**

Brood chamber
The part of the hive in which the **brood** is reared; may include one or more hive bodies and the combs within.

Brood diseases
Diseases that affect only the immature stages of bees, such as **American** or **European foulbrood.**

Brood nest
The part of the hive interior in which brood is reared.

Burr comb
Small pieces of comb made as connecting links between combs or between a frame and the hive itself.

Cage shipping
Also called a package, a screened box filled with 1–2.25kg (2–5lb) of bees, with or without a queen, and supplied with a feeder can; used to start a new colony, or to boost a weak one.

Capped brood
Immature bees whose cells have been sealed over with a brown wax cover by other worker bees.

Cappings
The name given to the thin wax covering over honey once it has been cut off the extracting frames; a source of premium beeswax.

Carnolian bees
A greyish race of honey bee, *Apis mellifera carnica.*

Cast
A second or subsequent **swarm** which leaves a colony with a virgin queen, after the first (or prime) swarm has departed.

Castes
The three types of bees that comprise the adult population of a honey bee colony: **workers, drones and queen.**

Cell
A single hexagonal compartment in a honeycomb.

Chalkbrood
A disease affecting bee larvae, caused by a fungus *Ascosphaera apis.* The larvae eventually turn into hard, chalky white 'mummies'.

Chilled brood
Immature bees that have died from exposure to cold; commonly caused by mismanagement.

Cluster
A large group of bees hanging together, one upon another.

Cocoon
A thin silk covering secreted by larval honey bees in their cells in preparation for pupation.

Colony
The aggregate of worker bees, drones, queen and developing brood living together as a unit in a hive or other dwelling.

Comb
The wax portion of a colony in which eggs are laid and honey and pollen are stored.

Comb honey
Honey in the wax combs, usually produced and sold separately.

Creamed honey
Honey that has been pasteurized and undergone controlled granulation to produce a finely textured candied honey which spreads easily.

Cross-pollination
The transfer of pollen from the anther of one flower to the stigma of another of the same species.

Dividing
Separating a colony to form two or more units.

Drawn comb
Combs with cells built out by honey bees from a sheet of foundation.

Drifting
The movement of bees that have lost their location and enter other hives.

Drone
The male honey bee which comes from an unfertilized egg.

Drone brood or drone comb
Brood which matures into drones, reared in cells larger than those of worker brood.

Drone congregating area (DCA)
A specific area to which drones fly waiting for virgin queens to pass by; it is not known how or when they are formed, but drones return to the same spots year after year.

Drone-laying queen
A queen who can lay only unfertilized eggs, due to age, improper or no mating, disease or injury.

Dwindling
The rapid dying-off of old bees in the spring; sometimes called spring dwindling or disappearing disease.

Dysentery
An abnormal condition of adult bees characterized by severe diarrhoea and usually caused by starvation, low-quality food and/or moist surroundings.

Eggs
The first phase in the bee life cycle enclosed in a flexible shell or chorion.

Entrance block
A notched wooden strip used to regulate the size of the entrance to a hive.

European foulbrood (EFB)
An infectious **brood disease** of honey bees caused by the bacterium *Streptococcus pluton.*

European honey bee
See *Apis mellifera.*

Extractor
A centrifugal force machine to throw out honey but leave the combs intact. May be either radial (working on both sides at once) or tangential (you have to turn it).

Feeders
Various types of appliances for feeding bees artificially.

Fermenting honey
Honey which contains too much water (greater than 19 per cent) in which a chemical breakdown of the sugars takes place, producing carbon dioxide and alcohol.

Field bees
Another name for **foragers.**

Flight path
Usually refers to the direction in which bees fly when leaving their colony; if the path is obstructed, the bees may become aggravated.

Forage
Natural food source of bees (nectar and pollen) from flowers.

Foragers
Worker bees which are usually 21 or more days old and work outside to collect nectar, pollen, water and propolis.

Foundation, wax
Thin sheets of beeswax embossed with the base of a worker cell on which bees will construct a complete comb (**drawn comb**).

Foundation, wired
Comb foundation which includes wires for added support.

Frame
Four pieces of wood forming a rectangle designed to hold honeycomb.

Fume board
A rectangular frame, the size of a super, covered with an absorbent material such as burlap, on which is placed a chemical repellent to drive the bees out of supers for honey removal.

Guard bees
Worker bees about three weeks old who guard the entrance to the hive.

Hive
A manmade home for bees.

Hive body
A wooden box containing frames.

Hive stand
A structure serving as a base support for a beehive; it extends the life of the bottom board by keeping it off damp ground.

Hive tool
A flat metal device with a curved scraping surface at one end and a flat blade at the other; used to open hives and to prise apart and scrape frames.

Honey
A sweet viscid material produced by bees from the nectar of flowers, composed largely of a mixture of dextrose and levulose dissolved in about 19 per cent water; contains small amounts of sucrose, mineral matter, vitamins, proteins and enzymes.

Honeydew
An excreted material from insects in the order Homoptera (aphids), which feed on plant sap.

Honey colour
Measured by a Pfund grader, honey colours are classified into seven gradations from water white and white to amber and dark amber.

Honey flow
A time when nectar-bearing plants are blooming, allowing bees to store a surplus of honey.

Honey sac
The part of the body in which bees carry nectar.

Honey supers
Hive bodies used for honey production.

Inner cover
An insulating cover fitting on top of the top super but underneath the outer cover, with an oblong hole in the centre.

Inverts
An enzyme in honey, which splits the sucrose molecule (a

disaccharide) into its two components dextrose and laevulose (monosaccharides).

Italian bees
A common race of honey bees, *Apis mellifera ligustica.*

Larva, capped
The second developmental stage of a bee, ready to pupate or spin its cocoon (about ten days after its emergence from the egg).

Laying workers
Worker bees which lay eggs in a colony that is hopelessly queenless; since the workers cannot mate such eggs are infertile and therefore produce drones.

Mandibles
The jaws of an insect; used by bees to form honeycomb, to scrape pollen, in fighting and in picking up hive debris.

Mating flight
The flight taken by a virgin queen during which she mates in the air with several drones.

Mead
Honey wine.

Metamorphosis
The three stages through which a bee passes before reaching maturity: **egg, larva** and **pupa.**

Migratory beekeeping
The moving of colonies of bees from one locality to another during a single season to take advantage of two or more **honey flows**.

Moisture content
In honey, the percentage of water should be no more than 18.6; anything higher than that will allow honey to ferment.

Movable frames
A frame constructed in such a way as to preserve the **bee space** and to be easily removed without damaging the bees; when in place, it remains unattached to its surroundings.

Natural honey
Unfiltered and unheated honey.

Nectar
A liquid rich in sugars, manufactured by plants and secreted by nectar glands in or near flowers; the raw material for honey.

Nectar glands
Special nectar-secreting glands usually found in flowers, whose function is to attract pollinating insects such as honey bees for the purpose of cross-pollination, by offering a carbohydrate-rich food.

Nosema disease
A widespread adult bee disease caused by a one-celled, spore-forming organism, *Nosema apis*; it infects the bee's gut lining.

Nuc, nuclei, nucleus
A small colony of bees often used in queen-rearing or as a 'starter' hive.

Nurse bees
Young bees, three to ten days old, which feed and take care of developing **brood.**

Outer cover
The last cover that fits over a hive to protect it from rain; the two most common kinds are telescoping and migratory covers.

Package bees
See **Cage shipping**.

Pheromone
A substance secreted by an animal which causes a specific reaction, such as stimulation to mate with or supply food to an individual of the same species.

Piping
A series of sounds made by a queen, frequently before she emerges from her cell.

Play flights
Short flights taken in front and in the vicinity of the hive by young bees to help them familiarize themselves with its location.

Poison sac
Large oval sac containing venom and attached to the anterior (front) part of the **sting**; stores venom produced by the poison gland; its primary ingredients are peptide and mellitin.

Pollen
The dust-like male reproductive cells (gametophytes) of flowers.

Pollination
The transfer of pollen from the anthers to the stigma of flowers.

Proboscis
The mouthparts of the bee that form the sucking tube or tongue.

Propolis
Plant resins collected and modified by bees, used to fill in small spaces inside the hive.

Pupa
The third stage in the development of the bee during which it is inactive and sealed in its cocoon.

Queen
A fully developed and mated female bee, responsible for all the egg-laying of a colony and recognized by other bees because of her special **pheromones**.

Queen cage
A special cage in which queens are shipped and/or introduced to a colony, usually with five or six young workers called attendants, and a candy plug.

Queen cell
A special elongated cell resembling a peanut shell in which the queen is reared; usually over 2.5cm (1in) in length, it hangs vertically from the comb.

Queen cup
A cup-shaped cell hanging vertically from the comb, but containing no egg.

Queen excluder
A device which permits workers to pass, but excludes the larger queens and drones.

Queenright
A colony that contains a laying queen.

Radial extractor
See **Extractor**.

Requeen
To introduce a new queen to a queenless hive.

Rendering
The process of melting combs and **cappings** and removing refuse from the wax.

Robbing
The act of bees stealing honey/nectar from other colonies; also applied to bees cleaning out wet **supers** or **cappings** left uncovered by beekeepers.

Ropy characteristic
A diagnostic test for **American foulbrood** in which the decayed larvae form an elastic rope when drawn out with a toothpick or matchstick.

Royal jelly
A highly nutritious, milky white glandular secretion of young (nurse) bees; used to feed the queen and young larvae.

Sacbrood
A **brood disease** of bees caused by a filterable virus which interferes with the moulting process; the dead larva resembles a bag of fluid.

Scout bees
Worker bees searching for a new source of pollen, nectar, propolis, water, or a new home for a swarm of bees.

Sealed brood
See **Capped brood.**

Self-spacing frames
Frames constructed so that they are a **bee space** apart when pushed together in a hive body.

Settling tank
A large-capacity container used to settle extracted honey.

Skep
A beehive without movable frames.

Smoker
A metal container with attached bellows which burns organic fuels to generate smoke.

Solar wax melter or extractor
A glass-covered insulated box used to melt wax from combs and **cappings** using the heat of the sun.

Species of bees
The four most common species of *Apis* are *A. mellifera, A. cerana, A. dorsata* and *A. florae.* Most of our garden bees are races (subspecies) of *A. mellifera* and other newly discovered races are currently under investigation.

Spermatheca
A small sac in which a queen bee stores spermatozoa.

Split
To divide a colony for the purpose of increasing the number of hives.

Sting
An organ belonging exclusively to female insects developed from egg-laying mechanisms, used to defend the colony; modified into a piercing shaft through which venom is injected.

Sugar syrup
Feed for bees, containing sucrose or table (cane) sugar and hot water in various ratios.

Super
A receptacle in which bees store honey; usually placed over or above the **brood nest**; so-called brood supers contain **brood**.

Supersedure
Rearing a new queen to replace the mother queen in the same hive; shortly after the daughter queen begins to lay eggs, the mother queen disappears.

Swarm
A collection of bees, containing at least one queen, that has split apart from the mother colony to establish a new one; a natural method of propagation of honey bees.

Swarm cell
Queen cells usually found on the bottom of the combs before swarming.

Tangential extractor
See **Extractor**.

Thorax
The central region of an insect to which the wings and legs are attached.

Travelling screen
A framed screen that fits over the top as a hive cover, used when moving bees in hot weather to provide sufficient ventilation to keep bees from suffocating.

Uncapping fork
A fork-like device used to remove wax cappings covering honey, so it can be extracted.

Uncapping knife
A knife used as an alternative to an **uncapping fork**.

Uncapping tank
A container over which frames of honey are uncapped; usually strains out the honey, which is then collected.

Uniting
Combining two or more colonies to form a larger colony.

Varroa jacobsoni
An external mite parasite on honey bees.

Veil
A protective netting that covers the face and neck; allows ventilation, easy movement and good vision.

Virgin queen
An unmated queen bee.

Wasp
A close relative of honey bees, usually in the family Vespidae; they are carnivorous, some species preying on bees.

Wax glands
Eight glands located on abdomen of young worker bees; they secrete beeswax droplets.

Wax moth
Usually refers to the greater wax moth, *Galleria mellonella*, whose larvae bore through and destroy honeycomb as they eat out its impurities.

Western honey bee
See *Apis mellifera*.

Winter cluster
A tight ball of bees within the hive which they form when outside temperature falls below 14°C (57°F) and they need to generate heat.

Worker bees
Infertile female bee whose reproductive organs are only partially developed, responsible for carrying out all the routine of the colony.

Worker comb
Comb measuring about two cells to the centimetre (five to the inch), in which workers are reared and honey and pollen are stored.

Index

Picture Credits

Brian McCallum 3c; Corbis 3L; Alamy 3R; Brian McCallum 7; Alamy 8–9; Bridgeman Art Library/Bibliotheque Municpale 10; 11; Alamy 12TR; Brian McCallum 12bl; Kim Taylor/The Garden Collection 13BL; Kim Taylor/The Garden Collection 14T, B; Brian McCallum 19; Jill Mead 20, 21l; Corbis 21R; Alamy 22; Jill Mead 23; Corbis 24; Alamy 27; Brian McCallum 27c; Nicola Stocken Tomkins/The Garden Collection 28–29 (Design: Cheryl Waller); E. H Thorne 32, 33, 34; Brian McCallum 38t, b, 47, 51; Alamy 51BR, 52–53; Brian McCallum 60; Alamy 61B; Liz Eddison/The Garden Collection 64–65 (Design: Patrick McCann); Neil Sutherland/The Garden Collection 66; Derek Harris/The Garden Collection 67; Torie Chugg/The Garden Collection 68; Corbis 69; Liz Eddison/The Garden Collection 70–71 (Design: Lucy Hunter); Alamy 74, 75R, 77; Jill Mead 78; Alamy 79; Brian McCallum 83, 85; Corbis 88; Bridgeman Art Library/Giraudon 89T, B; Corbis 90–91; Jonathan Buckley/The Garden Collection 94 (Design: David & Mavid Seeney); Jonathan Buckley/The Garden Collection 96–97 (Design: Helen Yemm, Ketley's); Brian McCallum 99b; Alamy 99T; Brian McCallum 101t; Alamy 101, 102, 107. 108; Getty Images 109; Punchstock 112; Alamy 113, 114L, R, 115L, R, 116L, iStockphoto 116R Alamy 117, Brian McCallum 119; Alamy 120; Brian McCallum 121.

Useful addresses and links

NATIONAL BODIES

Africa
The African Beekeeping Resource Centre
www.apiconsult.com/

Australia
Australian Honey Bee Industry Council
www.honeybee.org.au/

Canada
Canadian Honey Council
www.honeycouncil.ca
Canadian Association of Professional Apiculturists
www.capabees.ca

France
Union Nationale de L'Apiculture Francaise
www.unaf.net

Ireland
The Federation of Irish Beekeepers' Associations
www.irishbeekeeping.ie/

Italy
Istituto Nazionale di Apicoltura
www.inapicoltura.org

New Zealand
National Beekeepers Association of New Zealand
www.nba.org.nz/

Norway
Norges Birøkterlag
www.norges-birokterlag.no/

South Africa
Southerns Beekeeping Association
http://www.beekeepers.co.za/

USA
American Beekeeping Federation,
P.O. Box 1337, Jesup, GA 31598-1038
Tel: (912) 427-4233 Fax: (912) 427-8447
Website abfnet.org/

Sweden
Sveriges Biodlares Riksförbund
www.biodlarna.se/

UK
BBKA
BBKA HQ National Beekeeping Centre, National
Agricultural Centre, Stoneleigh Park, Warwickshire
CV8 2LG
Tel: 02476 696679 Fax: 02476 690682.
Website www.bbka.org.uk/
IBRA
International Bee Research Association
16 North Road, Cardiff CF10 3DY

www.ibra.org.uk/
National Bee Unit
Central Science Laboratory, Sand Hutton, York YO41
1LZ
Tel: 01904 462510
beebase.csl.gov.uk/index.cfm

MAGAZINES

New Zealand
New Zealand Beekeeper Magazine
http://www.nba.org.nz/Sections-article7-p1.htm

UK
Journal of Apicultural Research
http://www.ibra.org.uk/jar.html
BBKA News
http://www.bbka.org.uk/articles/bbka-news.php
Bee Craft
www.bee-craft.com/

USA
Bee Culture
www.beeculture.com/
American Bee Journal
www.dadant.com/journal

EQUIPMENT AND SUPPLIES

Australia
Bindaree Bee Supplies
http://www.bindaree.com.au
HoneyBee Australis
http://www.honeybee.com.au

Canada
Benson Bee Supplies
http://www.bensonbee.com
FWJones & Son Ltd
http://www.fwjones.com/
The Bee Works
http://www.beeworks.com/

USA
Betterbee
http://www.betterbee.com/
Arnold Honeybee Services
http://www.arnoldhoneybeeservices.com/

Denmark
Swienty A/S - Beekeeping Equipment
http://www.swienty.com/

South Africa
Southerns Beekeeping Association
http://www.beekeepers.co.za/

New Zealand
Ecroyd Beekeeping Supplies Ltd
http://www.ecroyd.com/

Ceracell Beekeeping Supplies Ltd
http://www.ceracellbees.co.nz/

UK
E.H.Thorne (Beehives) Ltd
http://www.thorne.co.uk
National Bee Supplies
http://www.beekeeping.co.uk/
Maisemore Apiaries
http://www.bees-online.co.uk/
Park Beekeeping Supplies
http://www.parkbeekeeping.com
Stamfordham Ltd
http://www.stamfordham.biz/
Kembles Bee Supplies
http://www.kemble-bees.com/

CHARITIES WORKING WITH DEVELOPING COUNTRIES

Bees for Development
http://www.beesfordevelopment.org/
Bees Abroad
http://www.beesabroad.org.uk/

Other useful links
Bees for Kids
http://www.bees4kids.org.uk/
Apiservices
http://www.beekeeping.com

Acknowledgments

The authors would like to thank Chris Deaves at the Twickenham and Thames Valley Beekeepers' Association for answering Brian's countless questions; John Chapple, chair of the London Beekeepers' Association for helping Alison with her first colony; John Hamer at Blackhorse Apiaries for enabling us to expand our apiary and knowledge; Jill Mead for her beautiful photographs of urban bees; our parents for all their never ending support and encouragement, including hosting some hives; friends for their fascination in our new-found hobby; and last but not least Mic Cady for chance meetings on trains and having faith in us to deliver this book.

The publishers would like to thank E. H. Thorne for their help with the hive illustrations, Ted Benton for his patience and contacts in the latter stages of the project and Lizzie Harper for doing the illustrations at such short notice.